Policing and Social Media

Policing and Social Media

Social Control in an Era of New Media

By Christopher J. Schneider

Foreword by David L. Altheide

LEXINGTON BOOKS
Lanham • Boulder • New York • London

Published by Lexington Books
An imprint of The Rowman & Littlefield Publishing Group, Inc.
4501 Forbes Boulevard, Suite 200, Lanham, Maryland 20706
www.rowman.com

Unit A, Whitacre Mews, 26-34 Stannary Street, London SE11 4AB

Parts of chapter 3 were previously published in Schneider, Christopher J. and Daviel
Trottier (2013), "Social Media and the 2011 Vancouver Riot," in Norman K. Denzin (ed)
40th Anniversary of Studies in Symbolic Interaction (Studies in Symbolic Interaction,
Vol. 40), Emerald Group Publishing Limited, pp. 335-362. Reprinted with permission of
Emerald Group Publishing Limited.

Parts of chapter 4 were previously published in Schneider, Christopher J., "Police Presentational Strategies on Twitter in Canada," Policing and Society: An International Journal
of Research and Policy June 4, 2014: Taylor and Francis, reprinted with permission of
Taylor and Francis Ltd, http://www.tandfonline.com.

British Library Cataloguing in Publication Information Available

Library of Congress Cataloging-in-Publication Data

Names: Schneider, Christopher J., author.
Title: Policing and social media : social control in an era of new media / by Christopher J. Schneider.
Description: Lanham : Lexington Books, [2016] | Includes bibliographical references and index.
Identifiers: LCCN 2016000115 (print) | LCCN 2016008835 (ebook) | ISBN 9781498533713 (cloth :
    alk. paper) | ISBN 9781498533737 (pbk. : alk. paper) | ISBN 9781498533720 (Electronic)
Subjects: LCSH: Police-community relations--Canada. | Police misconduct--Canada. | Social media--
    Canada. | Police and mass media--Canada. | Mass media and criminal justice--Canada.
Classification: LCC HV7936.P8 S38 2016 (print) | LCC HV7936.P8 (ebook) | DDC 659.2/
    936320971--dc23

Printed in the United States of America

This book is dedicated to my mother, Kathy, whose love knows no bounds and without whom my life and livelihood would not have been possible. I am forever grateful for her unconditional love, guidance, strength, and support. She is the wind beneath my wings.

Contents

Foreword

David L. Altheide

This significant book about police and social media affirms a much neglected proposition in sociological research: the social world is mediated, and events, activities, and institutional orders reflect the process by which this mediation occurs. Culture provides meanings and nuances that produce social order. The mass media and the information technologies and formats that transport and emphasize images, sounds, narratives, and meanings are crucial components of this meaning-making process. This may be referred to as a mediation process, which involves the construction and use of media logic to provide order and sense to the mass communication process that will be anticipated, understood, and shared by various audiences.

Developments in media studies, cultural studies, and communication research make it clear that information technologies and communication formats influence events, actions, and organizational routines. Notwithstanding the fine contributions by Peter Manning (Manning 2003b) and others about internal police communication practices, criminology, apart from cultural criminology (Ferrell et al. 2004), has been a bit slow in incorporating a broader theoretical perspective about the communication process into studies of policing, surveillance, and social control in general. Christopher Schneider's work is consistent with the path-breaking work of Richard V. Ericson, Aaron Doyle, and colleagues, who insightfully incorporated a mediated order perspective into major investigations of the interplay between the police and mass media in Canada. Schneider's lucid contribution traces and clarifies how media logic and social media formats have been incorporated into contemporary Canadian police practices. Using an innovative qualitative research design that captures major themes from Facebook, Twitter, and YouTube, Professor Schneider meticulously describes how institutional public police practices increasingly are being shaped by social media platforms. Case study materials of the 2011 riot in Vancouver also illuminate how social media contribute to collective behavior.

It is helpful to place this important study in the context of massive changes in information technologies that have profoundly affected many aspects of social life and, now, even police activities. While the political and cultural contexts played a significant role in these developments, it is also true that available communication technology—namely, print and television—gave a few institutional news sources the power to define situations and construct narratives of threat and disorder. Until recently, the police were the key information sources to news media about crime and disorder. Their unique relationship to defining the situation via the news media has been widely investigated, including several significant Canadian studies by Richard V. Ericson and his colleagues (Ericson, Baranek, and Chan 1987, 1989, 1991).

The public sense of disorder is in large measure due to decades of often distorted and sensationalized mass media reports from police sources about crime, danger, and mayhem (Kappeler and Potter 2005; Surette 2010). Particularly in the United States, this discourse of fear has traveled across topics such as street crime, drugs, random violence, missing/abducted children, illegal immigration, and, more recently, terrorism. The production of a culture of fear, often resulting in moral panics, has been widely chronicled (Altheide 2002; Furedi 1997; Glassner 1999; Goode and Ben-Yehuda 1994). With some exceptions, police agencies were the source of the material that fed the entertainment formats that have ruled commercial TV news reports since the 1970s. Fear became part of the entertainment format: dramatic, visual, emotionally engaging. Politicians and others took advantage of this baseline of fear by theatrically announcing a variety of efforts to protect the public from the source of this fear, usually street criminals. Proposals typically involved hiring more police and militarizing social control, expanding surveillance, and enacting draconian prison sentences and other punishments. Politicians in the United States mined this baseline of fear after the 9/11 attacks to expand surveillance and prosecute two wars in the Middle East that have ravaged the planet and advanced international terrorism (Altheide 2006).

The police continue to be, essentially, the "owners" of narratives about crime, fear, and danger, but this is changing as new information technologies and accompanying communication formats have empowered other organizations and individuals to capture visuals, videos, and provide instantaneous commentary, often disputing police claims. This change is Professor Schneider's main concern in this book.

Part of the lesson is that newer technologies (e.g., social media) free audiences from being merely recipients of information from one source (e.g., a TV newscast); just as citizens are now both objects and agents of surveillance, the mediated citizen sends and receives images formatted and framed by entertainment, brevity, visual interest, and fleeting temporal and spatial relevance. We are wedded to a theory of organizational truth grounded in immediate visual imagery. Most media users are grounded in entertainment logic that promotes idealized worldviews compatible with a variety of cultural narratives and ideologies, even if information sources include "fake news" or other sources (Wonneberger,

Schoenbach, and Meurs 2013). These considerations must be included as part of any theory of social change.

Christopher Schneider's book clarifies how some Canadian police agencies are adjusting to the digital revolution, when older technologies are less important than the pervasive social media. It is ironic that the police, who represent control and order, were rapidly drawn to the use of social media, whose users seem to represent freedom, spontaneity, and creativity. For example, in chapter 3 we see part of the process of social change through a creative comparison of police-media relationships in two Vancouver hockey riots—one in 1994 and another in 2011. An exhaustive study by Aaron Doyle (2003) of the 1994 riot illuminated how the police controlled the materials and a definition of the situation, and the nature and cause of the riot. The 2011 riot was different largely because of social media; thousands of citizens posted photos, videos, and comments about the riot participants, urging more citizen cooperation, and lauding the role of the police. Social media users were able to offer their own narratives of the disruptive behavior, with many shaming and retributive postings calling for tough sentencing. One officer commented that it was "a response Vancouver Police didn't anticipate, but had to adapt and respond to." Professor Schneider's discussion of the process and experience with several events, particularly the Vancouver hockey riot in 2011, illuminates how police officials adopted these media as a way to obtain citizen cooperation, gather evidence, and pursue an innovative mediated conversation with the public about crime, police work, and community involvement. This was a different "crime story" than had appeared with the 1994 riot. Schneider's account of how these actions, along with an evolution of personal Internet pages—ranging from MySpace to Facebook and others—can be read as a tightly focused chronicling of major institutional changes accompanying media formats.

One of the biggest changes of social media involves police–community relationships. This is now one of the most dynamic features of organized police activity. No longer are we stuck with official police edicts about "what happened." With video everywhere and an expanding number of police forces wearing body cameras, visual representations mitigate the police–community divide. A case study of the troublesome efforts of the Toronto police to manage the public perceptions of a video posted on YouTube of the killing of Sammy Yatim (chapter 5) illustrates the risk of poor adjustment to the new social media and the video-sharing platform YouTube. Viewed nearly two million times, the video generated intense discussion and commentary on police use of force.

The Vancouver riots, the Sammy Yatim killing, and other experiences have changed police accountability as social media expanded and as media use has become more personal, instantaneous, and visual. Many departments throughout Canada, the United States, and Europe are in the process of adding social media specialists. This promotes more surveillance. Moreover, Schneider documents how police organizations are adopting protocols to guide departmental communications as well as individual officers' comments as they use Facebook, Twitter, and other social media formats.

This ambitious book challenges researchers to continue to investigate the changing communications environment of police and social control agents. How social media will be accommodated and institutionally adopted to stringent police cultures remains to be seen. While the impact on police culture will always be contextualized by locale and unique social and historical circumstances, it appears likely that more police officers, especially a younger generation familiar with social media, will adapt to the new communications and accountability environment. Whether or not this alters police conduct remains an empirical question. Perhaps more important is the long-term impact on trust. As noted by Ericson and Haggerty, "Privacy, trust, surveillance, and risk management go hand in hand in policing the probabilities and possibilities of action" (1997, 6). While the long-term impact of these new technologies on police culture remains an open question, Professor Schneider has provided a lens through which to view it.

Preface

The emergence and proliferation of social media has been a catalyst for a profound seismic transformation in which the social construction of reality has shifted. This shift has contributed to new developments in social control, especially in relation to policing and police work. This book was in the final stages of revision when reports of a fatal encounter captured on video by a citizen bystander showed Los Angeles police officers shooting a man. Within hours, the story made international headline news. The citizen-recorded video, uploaded to social media, was viewed millions of times in less than a day. Reports indicate that two of the officers were wearing body cameras. The citizen-recorded video death of Eric Garner in New York and the death of Michael Brown in Ferguson, Missouri, each during police actions in 2014, gained considerable traction on social media and spurred various international protests. Readers familiar with coverage of these or similar events can find parallels with key points made in the following chapters.

This book aims to illustrate the process by which new information technology—namely, social media—and related changes in communication formats have affected the public face of policing and police work in Canada. I argue that *police use of social media has altered institutional public police practices in a manner that is consistent with the logic of social media platforms.* Policing is changing to include new ways of conditioning the public, cultivating self-promotion, and expanding social control. The global reach of new communication formats is evident, as are some related changes to police work elsewhere around the world, and for this reason international research projects on policing, police work, and cultural differences remain necessary.

This book is the culmination of over four years of research about social media and policing. It was the 2011 Vancouver riot in British Columbia, a new kind of mass-mediated criminal event that unfolded online as it did in real time, that piqued my interest in some of the topics covered in this book. Many very astute colleagues have contributed in various ways to this work. I am especially grateful to David Altheide for his very insightful and valuable suggestions on earlier versions of this manuscript. Gray Cavender, Kevin Haggerty, and Gary Marx provided many helpful suggestions on earlier drafts of chapter 3. My thinking has benefited from many conversations with good friends and colleagues Aaron Doyle, Jeff Ferrell, Wendy Gillis, Ariane Hanemaayer, Stacey Hannem,

John Johnson, Joe Kotarba, Paul Marck, Arthur McLuhan, Kyle Nolan, Marcia Oliver, Wes Pue, Carrie Sanders, Kevin Schneider, Christine Schreyer, Jessica Stites Mor, Dan Trottier, Peter Urmetzer, and Phillip Vannini. The work of the late Richard V. Ericson also influenced the direction and scope of this book. I am also grateful for the encouragement and support of numerous friends and colleagues at Brandon University, Wilfrid Laurier University, and the University of British Columbia. Many students in several courses, including "Police and Society," a course I have regularly taught over the last several years, helped me to think through many of the ideas in this book. Conversations with the following Canadian police officers also contributed to many of the ideas presented here: Deputy Commissioner Craig Callens, Commanding Officer, "E" Division of the Royal Canadian Mounted Police (RCMP); Detective Mark Fenton, Internet investigator of the Vancouver Police Department (VPD); Staff Sergeant Lawrence Smith and Detective Sergeant Cameron Field, both of the Toronto Police Service (TPS); RCMP police officers Superintendent Bill McKinnon, Corporal Sandi Fazan, Sergeant Ann Morrison, and Constables Mark Slade and Claudia Wytrwal.

Three anonymous peer reviewers of this manuscript provided many valuable suggestions for improvement. All shortcomings in this regard are my own. I wish to acknowledge Lexington Books acquisitions assistant editor Brighid Stone and copyeditor Joanne Muzak. Versions of two chapters have appeared elsewhere. A version of chapter 3, "Facebook and the 2011 Vancouver Riot," was originally published as Christopher J. Schneider and Daniel Trotter, "Social Media and the 2011 Vancouver Riot," Studies in Symbolic Interaction 40 (2013): 336–62. A version of chapter 4, "Police Presentational Strategies on Twitter," appeared as "Police Presentational Strategies on Twitter in Canada," Policing and Society: An International Journal of Research and Policy 26, 2 (2016): 129–47. I benefited from feedback provided on earlier drafts of these publications by anonymous peer reviewers as well as additional helpful suggestions by Judith Bishop, Norman Denzin, Jim Schneider, and Graeme Wynn. I am thankful to my family for their love and support. Thanks to Model Stranger for great music, and lastly thanks to my beloved English bulldog Tank for keeping me company while I wrote this book.

Introduction

Media Logic, Policing, and Social Media

This book investigates various public aspects of the management, use, and control of social media by police agencies in Canada. While each case study presented here focuses on a different social media platform or format, my concern is less with the particular format per se, as these will undoubtedly change, and more with developing suitable analytical and methodological approaches to understanding contemporary policing practices on social media sites in Canada. My analytical approach to police use of social media is media logic, and my methodological approach throughout the book is qualitative media analysis.

Media logic[1] is a theoretical model that examines "the process through which media present and transmit information" (Altheide and Snow 1979, 10). This book is concerned with police institutional orientation toward media presentation, specifically the most recent shift to social media. This shift to social media has seen policing expand to include new ways of conditioning the public and cultivating self-promotion. It has also brought about an expansion of social control efforts, defined, following Altheide, as *"the process by which people behave in ways that meet the expectations of others"* (2002, 13, emphasis original).

My approach in this book develops previous scholarship that has examined police institutional orientation toward the television medium (see figure 0.1) (Doyle 2003). Using media logic, I wish to illustrate how contemporary policing develops through use of social media, including altering police–public relations (see figure 0.2). As figure 0.1 shows, police modify institutional activities to develop new uses of oligopolistic media. The presentation of experience and information is carefully produced and selected to meet strategic objectives. Media materials are incorporated as a basic social form of routine elements of policing and police work. In figure 0.2, as police develop new uses of social media to meet institutional and strategic objectives, the presentation of experience and information is also carefully produced and selected to meet strategic objectives, but social media materials are more immediate and emotive than conventional media. Increased importance is placed upon police media performance, and mass media maintain oligopolistic functions.

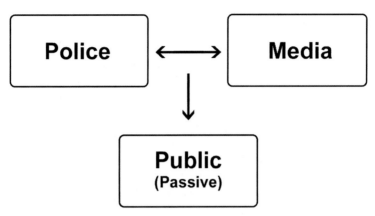

Figure 0.1.   Source: Christopher J. Schneider

Although scholarship in policing and social media is growing (see Lee and McGovern 2014), research in Canada has not explored how social media contributes to recent changes in police and policing. As sociologist Peter Manning notes, "the problems presented by the use of new data from social media, emails, videos, 'instant observations,' and data from ephemeral events that may be policed are yet to be faced" (2015, 273). Using qualitative media analysis (as discussed below), this book works with new data from social media to fill a gap in Canadian research on the role of social media in contemporary policing.

No universal consensus emerges from the case studies presented in this book; nor was it my intention to create consensus. Police are able to more effectively control and manage information in some circumstances than in others. The point, rather, is to show how attempts to make sense of events become quicker and *more* media-focused in order to illustrate the ongoing relations between social media and institutional police activity. Following news media, social media sites increasingly serve as important sources of legitimation in the social construction of reality, augmenting police visibility, as well as police legitimacy concerns (Chermak and Weiss 2005; Goldsmith 2010). The basic argument that this book advances is simple: *police use of social media has altered institutional public police practices in a manner that is consistent with the logic of social media platforms.* Logic here draws from Altheide and Snow (1979) and refers to the select format of social media sites and the manner that these formats influence how information is constituted, perceived, and then interpreted in online spaces. Format is the singularly most salient feature of media logic (Snow 1983). As Altheide and Snow explain, "Format consists, in part, of how material is organized, the style in which it is presented, the focus or emphasis on particular characteristics of behavior, and the grammar of media communication" (1979, 10).

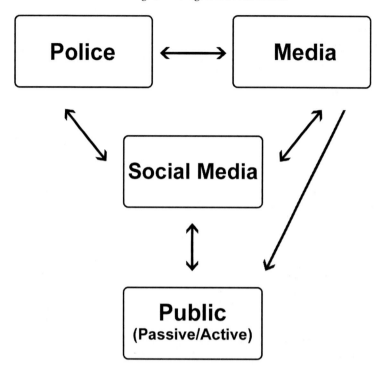

Figure 0.2.   Source: Christopher J. Schneider

These changes in police and policing have occurred very quickly over the last decade. This book traces some of these changes relative to a few key and significant historical moments, albeit those largely limited to Canada, including the launch of MySpace in 2003 and Facebook in 2004; the June 2011 Vancouver riot; the July 2011 launch of the Toronto Police Service (TPS) "social media strategy"; and the July 2013 bystander-recorded TPS shooting death of teenager Sammy Yatim.

Chapter 1 discusses media formats relative to police practices and outlines a few key theoretical perspectives. The second chapter begins with an investigation of police awareness of, and attraction to, MySpace and later Facebook, and connects this awareness to early police use of these sites primarily for investigative purposes. Chapter 3 illustrates how citizen-led justice efforts on Facebook after the riots in Vancouver in June 2011 contribute to the realization of the limitations of police control and definition of publicly circulated crime-related information and data. In the weeks following the riot, the TPS rolled out their social media strategy. Among the first of its kind in North America, it was intended as a large-scale, proactive, interactive media-oriented presentational strategy. Chapter 4 focuses on the TPS use of social media platform Twitter as a presentational strategy. The fifth chapter illustrates how the video-sharing site YouTube framed the police shooting death of a Toronto teenager ahead of police statements in news media. Each chapter demonstrates the ways that the particular logic of these formats collectively contributes to changes in institutional police practices, that taken together, helps to better situate and contextualize the transformative power of social media on policing and contemporary police work.

CONCEPTUAL ISSUES

There is no one precise definition of social media. One explanation for this ambiguity stems from the evolving and changing nature of social media. Definitions are not static. Not surprisingly, perhaps, an entire cottage industry of "social media professionals" has emerged in the wake of this definitional confusion. Many of these "professionals" are self-proclaimed "experts" on all things "social media." This includes social media law enforcement professionals discussed in chapter 4. No universal definition exists among social media professionals. To illustrate the point, a survey of these social media professionals reveals dozens of social media definitions (Cohen 2011). Confusion regarding a shared definition of social media has also been shown to exist among professional groups, such as university professors (Schneider 2014).

While difficult to define precisely, all forms of social media share a few basic characteristics: they enable creation, rely exclusively on audience participation relative to the production of content, and involve varying degrees of user engagement. For our purposes, social media can be understood as a hybrid of social interaction and media. This definition is intentionally flexible in order to include a variety of "technical and social practices" (Mandiberg 2012, 5). Conceptually, this definition also allows the phrase "social media" to be used interchangeably with a host of monikers that are also used in reference to this phenomenon, the most popular of which includes "social network."

In an influential definitional essay on the subject, boyd and Ellison define a social network as "web-based services that allow individuals to (1) construct a public or semi-public profile within a bounded system, (2) articulate a list of other users with whom they share a connection, and (3) view and traverse their list of connections and those made by others within the system. *The nature and nomenclature of these connections may vary from site to site*" (2007, 211, emphasis mine). The confluence of this phenomenon, whether it is called a social network, social media, or another name (e.g., new media), all share a fundamental characteristic: *the production of audience representations in the form of documents*. The audience *and* the content are each significant features of social media. Document production, then, is an important and definitive characteristic of social media. Documents "can be defined as any symbolic representation that can be recorded or retrieved for analysis" (Altheide and Schneider 2013, 5).

This might include text such as posts on social media, photographs, videos, or audio recordings. Much of this book focuses on text-based documents.

The study of documents is important because documents capture and reflect social meanings as well as institutional relations, including recognizable empirical changes to the institution of policing, as this book illustrates. From this core assumption about the nature of documents emerges a set of research questions and the determination of a suitable methodology. I use qualitative media analysis (QMA) as my methodology (see Altheide and Schneider 2013). Since the first edition of *Qualitative Media Analysis* (1996), numerous high-quality peer-reviewed journal articles, book chapters, master's theses, and doctoral dissertations have used this method.[2] QMA is also listed as an acceptable research method in the *Sage Encyclopedia of Social Science Research Methods* (Altheide 2004).

METHODOLOGY

Qualitative media analysis provides a framework to define, organize, and examine documents. This process allows the researcher to draw theoretical inferences from selected materials, such as data in the form of documents collected from social media. Research and awareness in the production of documents, such as posts made to social media platforms, helps inform the selection of sampling procedures. This process involves constant comparison and discovery to inform specific categories and narrative descriptions. Progressive theoretical sampling and saturation sampling were each used in the chapters that follow. A saturation sample was drawn by entering selected terms into the LexisNexis database in chapter 2 and later in each of the collected social media data sets. These terms are briefly outlined and discussed further in the Research Questions and Chapter Overview.

Progressive theoretical sampling includes the selection of materials drawn from the emerging understanding of a specific topic. This form of sampling ensures that the full range of materials is included. "The idea is to select materials for conceptual or theoretically relevant reasons" (Altheide and Schneider 2013, 56). The initial step in this process involves becoming familiar with the topic at hand—namely, police involvement with social media sites. Reading and learning about the organizational context of social networking sites (see boyd and Ellison 2007) such as Facebook, Twitter, and YouTube helps provide insight into the organization routines in the production and presentation of user-generated posts by citizens and police.[3] This initial work revealed the significance of the role of Facebook in the 2011 Vancouver riots, the Toronto Police Service's acknowledgment of the need to use Twitter, and the role of YouTube in the construction of the shooting death of Toronto teenager Sammy Yatim.

The integration of data materials relies on the researcher's identification of the meanings and contexts of documents. As Altheide and Schneider note,

> Qualitative data analysis is not about coding or counting, although these activities can be useful in some parts of fulfilling the goals of the quest for meaning and theoretical integration. . . . The goal is to understand the process, to see the process in the types and meanings of the documents under investigation, and to be able to associate the documents with conceptual and theoretical issues. This occurs as the researcher interacts with the document. [Therefore] it is best to rely on the more straightforward "search-find-replace" options on most word processing programs. (2013, 70)

The methodological nuances of the process of qualitative document analysis are discussed in greater detail in the second edition of *Qualitative Media Analysis* (see Altheide and Schneider 2013). Below, I outline questions specific to each chapter along with a short context of the methodological procedures in response to the research questions.

RESEARCH QUESTIONS AND CHAPTER OVERVIEW

Chapter 1 provides a broad overview of changes in media formats and police control practices in order to help illustrate developments of police strategies around social media practices. The hiring of social media officers is one example. As noted by deputy chief Peter Sloly of the TPS, social media "enables us to do old business in newer ways" (COPS and Police Executive Research Forum 2013, 5). The TPS hired the first social media officer in Canada in April 2010. These and other social media related strategic positions have since quickly expanded across police services. Other examples of police practices on social media are briefly noted and discussed. I then turn my attention to conceptualizing and defining pertinent theoretical concepts, including social control and media logic.

Chapter 2 asks, *how and when did police become involved with social media?* This chapter explores the police attraction to social media, specifically the social and historical circumstances that give rise to policing on social media. Most official police accounts on sites such as Facebook and Twitter started to appear with increased frequently between 2009 and 2011. The first social networking sites launched in 1997 (boyd and Ellison 2007), and hundreds of these sites existed for nearly a decade without an official public police presence. This is not meant to suggest the absence of police on social media, in either undercover or surveillance work, but, because this work is usually secretive, it can be difficult to detect (Schneider 2015a). In an effort to identify early police interest in these sites, even prior to the police's official use of them, it makes sense, then, to focus on relevant news documents. These news documents were located using the

LexisNexis database. Beginning with selected search terms "MySpace" (one of the first social media sites to be regularly associated in news media with crime, which helped to attract the attention of police) and "police" led to the discovery of other terms and phrases, including "sexual predators," "under-age," and "teenager." The use of these terms led to recognizable patterns and themes across examined news reports. The chapter illustrates the quick spread of sensationalized discourse that helped attract and retain the attention of various police agencies. Police use of these sites to aid investigations as discussed in news media accounts quickly followed and later expanded to other criminal and non-criminal matters, including recruitment and police presentational strategies. The rest of the chapters in the book explore some of these issues in more specific contexts.

Chapter 3 explores the "first North American social media sports riot" (VPD 2011, 7) and one of the first large-scale instances in Canada of crime in progress as it initially unfolded nearly in real time on social media: the 2011 Vancouver Stanley Cup riot. "Riots" are complex social events that most always involve police who officially designate these gatherings as riots. This designation allows police to respond in a variety of ways that includes the use of legitimate force. With the popularization of social media, however, these situations are now defined not only by police but also by citizens. Thus, citizen-driven definitions of the situation might also exist alongside official police definitions, which can complicate various police social control efforts (Longley 2013). This chapter asks, *what does the online meaning-making process in response to the 2011 Vancouver riot tell us about meaning making in public spaces in relation to criminal events such as riots? And what might user activity on social media tell us about the social process of the definition of the situation, including police control of the definition?*

A very similar riot occurred in Vancouver in 1994. Police were able to successfully control documentation of the 1994 riot and its subsequent interpretation by citizens (Doyle 2003; Thorbes 1994). The exact opposite occurred in 2011: police were initially unable to monopolize, frame, and control understanding of the riot (Furlong and Keefe 2011; Longley 2011, 2013; VPD 2011). Social media users were able to piece together narratives of the riot without the assistance of the police or the law (Schneider and Trottier 2012; Schneider 2015a, 2015c; Trottier 2012a). Collecting narrative data in the form of user posts on the social media site Facebook helped identify key themes and frames related to the online collective meaning-making process. Social media have greatly expanded sources of documents for the purposes of investigation (Grogan 2014; International Association of the Chiefs of Police 2013). Social media have also become a relevant source for developing understandings about audiences' views and the narratives that are associated with select problems like riots (Schneider 2015a, 2015c; Trottier 2012a).

In the weeks that followed the 2011 riot, 12,587 postings were made to the primary "wall" of the "Vancouver Riot Pics: Post Your Photos" Facebook group page—one of the most popular user-created and -controlled social media sites dedicated to the riot. The frame of the riot was set by the contours of the page itself as guided by its users, including what was, and was not, discussed on Facebook by users. Using search terms such as "riot," "guilty," "crime," and so on, each consistent with the frame of the event, I retrieved selected posts across an Adobe portable document format (PDF) data set of 2,118 pages of Facebook data. The accumulation of "definitions of situations" (Waller 1970) provides some insight into the process of collective user narratives and interpretations, the consequences of which can have implications on the nature of police work and the development of police strategies, including the police solicitation of evidence from social media users.

In what ways are police in Canada using Twitter? And how does the use of Twitter contribute to the overall development and expansion of police presentational strategies? Many police departments in Canada now have a social media presence. As chapter 4 explains, police department uniformity on social media in Canada was not fully implemented until July 2011 when the Toronto Police Service, building "upon a previously existing strategy of individual contacts with the media" (Meijer and Thaens 2013, 348), launched their "social media strategy" just weeks following the 2011 riot in Vancouver. Discussions of the TPS strategy initially occurred as early as 2010, however (Connected COPS 2011). Thus, the timing of the launch so close to aftermath of the Vancouver riot appears to be a coincidence.

The TPS is a recognized leader across North America for its use of social media (Longley 2013; Meijer and Thaens 2013; Murray 2012). While other agencies are now on social media, the TPS is the leading police agency on social media in Canada (Murray 2012). At the time of this writing, the TPS has more authorized accounts on Twitter than any other police force in Canada. Data from official TPS accounts on Twitter are relevant for understanding symbolic meanings and images associated with police presentational strategies. Directing our attention to the TPS is important because other police agencies in Canada have modeled their use of social media following the TPS strategy (Longley 2013). In the course of my research, I collected 105,801 tweets across 119 TPS-authorized Twitter accounts. These Twitter data were then combined into a 7,498-page searchable PDF document. Research on police use of Twitter has shown that police use involves the circulation of crime-related information (Heverin and Zach 2010). I entered search terms consistent with police work and police use of Twitter, such as "crime" (which appeared 6,631 times across these data), into the PDF data set. This process resulted in the discovery of two dominant themes on Twitter: police professionalism and community policing.

While chapters 2 and 3 are concerned with issues of police awareness of social media and chapter 4 looks at control and management of social media as a presentational strategy, chapter 5 explores how contemporary "crime stories" (Chermak 1995) that involve police can very quickly unfold on social media, ahead of official interpretation provided by police statements like those featured in news media. *How do crime stories that involve police that begin on social media impact the policing institution?* Chapter 5 reveals that attempts to make sense of situations unfold quicker online, where user debates rapidly emerge and understandings of socially constructed reality are more media-oriented. The chapter takes a somewhat different approach than other chapters in that it focuses on the *inability* of police to effectively control the circulation of materials and user debates on social media.

In July 2013, a bystander video recording of the police shooting death of Toronto teenager Sammy Yatim was uploaded to YouTube in the minutes that followed the shooting. This video became the point of reference for both police and news media. The discussion that ensued on the YouTube page that featured the original video of the shooting framed the death of Yatim and provided an online and interactive interpretative social context for the emergence of meanings associated with the shooting. In just one week, the video was viewed more than 600,000 times and 8,586 posts were made to the YouTube page. I collected these posts into an 826-page PDF document, and entered search terms such as "knife" and "weapon" into the searchable data set. This chapter illustrates how social media can frame crime stories entirely without police narratives—a process that ushers in changes to police work, including the incorporation of police body cameras to manage institutional messages more effectively (Brown 2013).

This book's conclusion summarizes main themes and findings and briefly offers several recommendations for future research on policing practices on social media. Further research is necessary to deepen our understanding of international police agencies' efforts to respond and adapt to social media. And we need to better understand how advancements in social media have impacted and altered police work.

While this book attempts to contribute generally to the scholarship of social media and policing, the specific focus is on Canadian policing, and even more specifically on two police departments in Canada. Therefore, it merely scratches the surface of developments in policing and social media. With some exceptions (e.g., Brown 2013, 2015; Goldsmith 2010, 924–26; Meijer and Thaens 2013; Milbrandt 2010, 128–29; Nolan 2014; Schneider 2015a, 2015b, 2015c; Schneider and Trottier 2012, 2013; Trottier 2012a, 2012b), only minimal work has examined developments in policing on social media in Canada, including how social media materials and publics on social media are managed, or not managed (chapter 3), by police agencies, recent developments in police image work strategies (chapter 4), and growing challenges to police legitimacy (chapter 5). Internationally, research in these areas has investigated contemporary developments in police media work on social media in Australia, the United Kingdom, and the United States (Lee and McGovern 2014).

There are some important limitations of this volume worth highlighting at the outset. First and foremost, the three case studies, data from social media sites, and conclusions that are presented in this book are not representative of police and police work as a whole, or even across Canada. Nor are these materials necessarily intended to be an exhaustive representation of police use of social media in Canada. Rather, the materials and case studies presented herein are intended to contribute more generally to the limited body of research that explores the impact of technology on police work (Koper, Lum, and Willis 2014). The point is not to extrapolate from these materials to make predictions about police work at the local, national, or even international levels. Second, while this book is mostly limited to public elements of Canadian police and policing on social media, research that analyzes how social media materials are used internally by police organizations remains necessary. This includes how social media create "new social objects" (Manning 2008) much beyond the scope of various police and user-generated materials presented in public online spaces as explored in this book. Much less is actually known about the internal police use of various social objects created and produced by social media for investigation purposes, evidence, and so on. A study on the social organization of policing following the work and innovative approaches of other scholars, most notably Peter Manning (1997), would contribute to our limited understanding of internal police use of these data for crime mapping and analysis and calls for service, for example (see also Manning 2008). This is another book that remains to be written.

It is an intellectually exciting time for those interested in understanding changes to the social order that include social control and the role that police play in affirming order through social media. In many ways, this book is a continuation of the decades of scholarship that has sought to develop our understandings of the relations between police and media (Ericson 1982; Fishman 1978, 1980; Reiner 2003; Skolnick and Fyfe 1993; Lee and McGovern 2014; Mawby 1999).

The literature on police and media has illustrated the manner in which police work has developed to include media as a core strategy. One basic contribution that this book makes to the expanding literature in this area is that it is driven largely by an empirical focus on public perception online in the form of user-generated content. Furthermore, with few exceptions (Lee and McGovern 2014, 124; Meijer and Thaens 2013), most of the research on the influence of social media on policing ignores how the logic of social media plays out in police practices. No research has systematically examined the incorporation of this logic into police practices in Canada. Thus, a few basic questions emerge: First, what social and historical contexts gave rise to the incorporation of social media into the modern institutional strategies of police? And, second, what are some of the consequences of this process, particularly in relation to social control efforts and the definition of the situation? These two questions help guide the direction, scope, and focus of this book.

NOTES

1. *Media logic* was coined and developed by Altheide and Snow (1979, 10). My use of media logic in this book builds on decades of scholarship (Altheide 2014) directed toward understanding connections between media formats and how institutions re-shape themselves to fit media needs.

2. See the appendix in Altheide and Schneider for a sample of these works (2013, 133–37).

3. Unless otherwise noted, all time stamps from user-generated posts cited in this book are in Pacific Standard Time, the time zone where these data were collected.

ONE

Media Formats and Police Social Control Practices

In 2013, in response to growing concerns about the spread of social media, the FBI National Executive Institute Associates (NEIA), a U.S. law enforcement organization that focuses on education, research, and training for law enforcement, reported, "the question is no longer whether the police will use SM [social media], it is just *how quickly* and *how well* we will do it!" (2013, 7, emphasis mine). Deputy Chief Peter Sloly of the Toronto Police Service, who worked on the NEIA publication *Social Media: A Valuable Tool with Risks*, noted that social media "enables us to do old business in newer ways" (COPS and Police Executive Research Forum 2013, 5).

Most law enforcement agencies across Canada, and international police agencies, would likely agree with these two assertions as social media strategies become embedded in departments worldwide. The first statement, made by the NEIA, appears in the aforementioned *Social Media: A Valuable Tool with Risks*, a 280-page report written collaboratively by chief executive officers of the largest law enforcement agencies in North America. Law enforcement agencies, including many police departments, are increasingly using various social media platforms as corporate communication strategies to interact with citizens as well as to disseminate all kinds of information (Meijer and Thaens 2013). According to Murray Lee and Alyce McGovern, authors of *Policing and Media: Public Relations, Simulations, and Communication*, "strategizing around social media has become a priority within media and public affairs branches of police organizations [and] as a result, platforms such as Twitter, YouTube, and Facebook have experienced growth in usage by police organizations eager to reap the rewards of carefully planned social media agendas" (2014, 115). Police social media agendas can be best understood in relation to three interrelated "rationalities that underlie modes of media and public engagement by police": *publics, police image,* and *police legitimacy* (Lee and McGovern 2014, 38). This book aims to better understand and clarify related developments in these areas by focusing on two large Canadian police departments and their use of Facebook, Twitter, and YouTube. The police practices discussed herein involve the public face of police engagement on social media, a phenomenon that has spread rapidly across the globe (Lee and McGovern 2014). Public engagement on social media illustrates the concept of "new visibility" (Thompson 2005)—the idea that "making visible of actions and events is not just the outcome of leakage in systems of communication and information flow that are increasingly difficult to control; it is also an explicit strategy of individuals who know very well that mediated visibility can be a weapon in the struggles they wage in their day-to-day lives" (Thompson 2005, 31). Previous work has applied this concept to policing, noting that this visibility is "a critical component of how they *appear to the public* [and therefore] invites assessments of the propriety (or rectitude) of their behavior and thus plays an important role in determining public reactions to the police (Tyler 2005)" (Goldsmith 2010, 914, emphasis original). Since police departments, too, are subject to this visibility, police are using social media to address concerns over "corporate reputation" (Lee and McGovern 2013), for example.

Beyond the promise of effectiveness of the perceived police use of social media, little is known about the impact of social media on police and policing, including the possible long-term implications that police use of social media may have for police legitimacy concerns (Meijer and Thaens 2013). Wesley Skogan and Kathleen Frydl define police legitimacy as "the judgments that ordinary citizens make about the rightfulness of police conduct" (2004, 291). As Albert Reiss argues, "the way police *exercise* their authority in encounters with citizens is important in maintaining its legitimacy" (1971, 175, emphasis original). Police legitimacy, including public perceptions of "fair procedures," rests on "the hearts and minds of the public" (Tyler 2004, 91). These perceptions "are created and sustained by the process of policing itself" (Skogan and Frydl 2004, 291–92), and this process is what entitles police to be obeyed by citizens as legitimate authorities (Reiss 1971; Tyler 2004). Research suggests that effective order maintenance and police legitimacy relies on media (Ericson, Baranek, and Chan 1989, 1991; Mawby 2002b). People learn about law and policing, including police procedures, largely from exposure to media, rather than direct experience (Macauley 1987; Schneider 2012).

POLICE IMAGE WORK

A significant part of police exposure includes police control over their public perception via the media as the legitimate authority (Ericson 1982; Mawby 2002b). The control and use of social media for perception purposes includes "image work" (Ericson 1982, 10) as a basic feature of the police mandate (Manning 1978). Image work refers to "all the activities in which police forces engage . . . [to] project meanings of policing" (Mawby 2002b, 1), including "repair work" (Goldsmith 2010) and police "presentational strategies" (Manning 1978). Police image work has taken various forms; what has remained constant, however, is that this work has always existed in relation to the legitimation process (Mawby 2002b).

Police image work is an ongoing area of scholarly investigation. Previous research has explored the impact of technology on policing (Chan 2001), but this research remains underdeveloped (Koper, Lum, and Willis 2014). Nevertheless, work in this area has added to our knowledge of technological developments in police and policing in North America (Meijer and Thaens 2013), the United States, Britain (Manning 2003a, 2008); Leishman and Mason 2003), and Australia (Chan 2001; Lee and McGovern 2014). However, much less is known about the impact of social media on police and policing in Canada, which this book aims to help remedy.

SOCIAL MEDIA OFFICERS

Policing efforts on social media now include the generation of official police information that is not directly associated with popular understandings of police work, or crime work (Ericson 1982). Policing agencies around the globe are now recruiting "social media officers" to operate and manage official police social media accounts. These recent developments have even occurred ahead of police department policies on social media (e.g., VPD 2012). Although very little is known about the effectiveness and impact of social media technology on policing, "the current emphasis on police technology reflects a strong belief (among both police and citizens) in its potential to enhance policing" (Koper, Lum, and Willis 2014, 213).

In 2007, the Toronto Police Service (TPS) became the first Canadian police agency to use social media on a "structural basis" (Meijer and Thaens 2013, 347). In April 2010, Constable Scott Mills of the TPS became the first officer in Canada to hold an official police title as social media officer. Social media officers are sworn police members who have all the same duties and responsibilities as other officers of the law, but they are also responsible for managing the department's social media presence. As early as "2007, Constable Mills began to feel that the TPS was 'missing the boat' on social media" (COPS and Police Executive Research Forum 2013, 4). Following the hiring of Constable Mills as social media officer, various job postings of this kind and demand for officers fluent in these technologies have continued to increase across Canada. A 2014 Regina Saskatchewan Police Services application for a social media officer illustrates the point. The description calls for applicants to have an "understanding of portal, collaborative and social networking technologies, hand-held communication devices, etc., and the ability to troubleshoot on website and social media technologies."

In Texas, in February 2011, the Dallas Police Department hired Corporal Melinda Gutierrez as their social media officer. Reflecting on her position, Corporal Gutierrez writes, "The department was joining the 21st century and my job was to help get us there" (2011). Describing her new officer duties, she continues,

> Our Facebook, Nixle,[1] and one of the Twitters [sic] accounts are public sites that are open to the community. I am continuously posting information on our Facebook page which is in turn linked to our public Twitter account. This allows me to get information out on both simultaneously. Some examples of items that get posted on our page include: surveillance videos of cases where our homicide or robbery units are asking for the public's help, department-sponsored events and fundraisers, award announcements, and press releases. I use our "Notes" page to post answers to questions most commonly asked, from "How to Commend an Officer" to "Obtaining Offense and Accident Reports." Also I created a takedown policy for inappropriate posts to our Facebook accounts and placed it on our information page. This informs the community about what types of comment and photos will not be permitted on our page. One of our Twitter accounts is for departmental use only. This page was set up so that I could get departmental information out to police personnel rapidly and efficiently. Items usually sent out on this account range from retirement and award announcements, fundraisers, and information regarding sick or injured officers. (2011)

Social media officers are an outgrowth of police communication officers. The primary task of these emergent positions is to develop social media in support of department priorities. This includes, among other tasks, daily maintenance of department social media accounts.

SOCIAL MEDIA AND THE REPRODUCTION OF SOCIAL ORDER

Symbolic communication and control of information is a basic element of police work (Manning 1997), the explicit purpose of which is to affirm and reproduce social order (Ericson 1982). This includes the control and management of media sources (Ericson and Haggerty 1997) to ensure public compliance to police authority. Traditionally, these processes have occurred mainly through news media (Fishman 1978, 1980) and enter-tainment media (Doyle 2003; Fishman and Cavender 1998). While enter-tainment media can contribute to shared understandings of what it is that police do and what police work entails (Macauley 1987; Schneider 2012), news media remains "pivotal" to police as authorities must "make con-vincing claims" to the public (Ericson, Baranek, and Chan 1989, 8). This claims-making process derives from police reporting practices of crime (Fishman 1978). Thus, the question emerges: How are police using social media to reproduce social order?

Social media platforms are the most recent addition to the media land-scape. On these platforms, information in the form of visual, audio, and text can be viewed and heard on demand by most anyone anywhere. This includes both *authorized* and *unauthorized* information created by police as well as various materials generated by citizens. Social media augment institutional authority concerns of control of an ever-expanding array of media sources (Trottier 2012b). The institutional shift toward specialized police officers tasked with management of official social media accounts attests to some of these concerns. As additional evidence of this shift, one might take into account a 2014 statement by the York Regional Police in Ontario in the "Media Contacts" section of their website. This excerpt signals the explicit police acknowledgment of the need to manage social media:

> The York Regional Police Media Relations Officers are the public voice of York Regional Police. In 2012, we issued more than 600 media re-leases on subjects ranging from police activity to crime prevention and personal safety messaging. Our team also attends community events and the scenes of major incidents to liaise with the media and ensure our citizens receive accurate and timely information. *Our Media Rela-tions Officers also work in conjunction with the Corporate Communications team to manage our presence on social media.* (Emphasis mine)

As of 2015, the York Regional Police do not have a designated social media officer. Instead, their media relations have merged with the man-agement of social media platforms.

The York Regional Police joined Twitter in January 2011. A total of 7,396 "releases" in the form of published posts (known as tweets) have been made to the account over approximately forty months, far surpass-ing even a conservative estimate of standard police news media releases during a comparable period (normally, at least six hundred releases are produced annually). In some instances, there is overlap between releases to news media and social media. Police postings on social media usually contain far less content than their news release counterparts. Content that appears on social media is usually concise—select words (e.g., names, dates, location), pictures, and, most often, links to the more detailed news media releases that redirect the user to the York Regional Police website.

On the one hand, social media helps improve the control of the dissemination of information; on the other hand, there is a new public expectation that police will populate social media platforms with a consistent stream of official information, including information that is not directly relevant to police or crime work (Ericson 1982).

For instance, police agencies discuss their need to use their social media accounts to highlight their belief in the importance of social media via these same accounts. Such claims are indicative of the introduction of social media technology into police developing strategies. It would certainly be quite bizarre to watch a televised news media conference or read a newspaper report of police agencies highlighting the importance of using these media simply for the sake of using media absent any sort of context (e.g., crime stories). Consider again the York Regional Police.

On April 28, 2014, the York Regional Police tweeted, "Social Media training has become a significant component of the Media Relations course. SM is a key communications tool for law enforcement" (April 28, 2014, 5:04 a.m.). There are numerous posts on Twitter just like this one, offered by various police agencies. The point is that police strategies continue to evolve in response to changing media as formats expand from oligopolistic (i.e., "traditional" news media) to more interactive media (i.e., social media).

Public police use of social media is more than information dissemination. Police not only distribute authorized information and interact with citizens on social media platforms; they also use these interactive spaces to manage citizen-generated information that might reflect poorly upon police.

SOCIAL CONTROL

As Michael Banton wrote in his influential 1964 book *The Policeman in the Community*, "The police are only one among many agencies of social control" (1). Although police "must be seen as a specific aspect of [social control]" (Reiner 2010, 5), there are many forces of social control. However, no standard, precise, or universally agreed upon definition of social control exists in the sociological research literature (Meier 1982). Most broadly, social control is "the behavior of any individual that influences the behavior of others" (Gibbs 1977, 441). Also generically, social control refers to "the capacity of society to regulate itself according to desired principles and values" (Janowitz 1991, 73). As Albert Hunter explains, "The generic conception [of social control] is analytically more powerful [than specific definitions] for it allows one to distinguish among different forms of social control, of which coercion is only one, and also to account for historical and situational variations in the forms of social control" (1985, 231). These variations may involve formal and informal classification and regulation practices (Cohen 1985). The regulation of behavior involves setting the parameters of acceptable conduct and defining social situations and norms (Goffman 1959), including the relation of such norms to the capacity of self-regulation among social groups (Janowitz 1978).

Informal agents of social control, such as parents and teachers, govern conduct by enforcing rules, norms, and social values by applying sanctions in response to violations. In more complex social orders—for example, where group members represent different and sometimes competing social interests—formal social control develops in the form of law as a specific mechanism to preserve social order and social solidarity (Durkheim 1966). As Tom Van den Broeck explains, "The effectiveness of formal social control is in many ways determined by informal social control [and this relation presents a dilemma in that] the decline of informal social control causes a demand for more formal social control" (2012, 210; see also Jones and Newburn 2002). An essential feature of the development of contemporary social control involves social responses to crime (Durkheim 1982), which includes enforcement of criminal law as a "conflict management" strategy by third-party, state-controlled, and -sanctioned agents—for example, the police—which, notably, may include state-authorized (i.e., "legitimate") use of force (Black 1983; Bittner 1970; Hunter 1985; Van den Broeck 2012). According to Donald Black, "Much of the conduct described by anthropologists as conflict management, social control, or even law in tribal and other traditional societies is regarded as crime in modern societies" (1983, 34). Criminal law develops as an essential element of formal social control in societies where other traditional forms of control have failed (Manning 1978).

Most of the time, police do not use criminal law to restore order, and they relatively rarely make arrests (Bayley 1994). However, police painstakingly work to maintain the appearance that they spend all of their time using criminal law to restore order, including by making arrests (Manning 1978). The dramaturgical[2] organization of police work (Manning 1997),[3] particularly in relation to crime and law, develops in accordance with public expectations that "real police work is crime work" (Ericson 1982, 5). An important aspect of "real police work," therefore, is the ability to successfully maintain symbolic control over crime matters (Manning 1978, 1997). In short, "the police [maintain a] monopoly on public discourse about crime" (Kasinsky 1995, 212). Much of this symbolic work is accomplished through explicit promotion in news media (Brodeur 2010; see also Doyle 2003; Ericson, Baranek, and Chan 1989, 1991; Fishman 1980; Mawby 1999). Nevertheless, the active relations between media, police, and social control are often overlooked even today, at a time when police agencies around the world are increasing their presence and activity on various social media platforms.

As Jean-Paul Brodeur explains, "The role of media in social control has generally been examined from the standpoint of their contribution to the definition of deviance and the part they play in generating moral panics (Critcher 2003). *Their contribution is actually much more concrete: in several aspects of their functioning, the media are an actual and active part of the police assemblage*" (2010, 85, emphasis mine). "The task of empirical social research," in Morris Janowitz's words, "is to investigate the forms and consequences of social control" (1991, 75). Media definition of social situations is one source—albeit a powerful one—of social control (Ericson, Baranek, and Chan 1989; Ericson 1991). Very little research has explored how this process is enacted via police use of social media. This book addresses this gap in research by focusing squarely on social media as an active part of the police assemblage.

Social control is defined here, following David Altheide, as "*the process by which people behave in ways that meet the expectations of others*" (2002, 13, emphasis mine). This process is occurring with increased frequency on social media. According to Janowitz, "the problem is whether the processes of social control are able to maintain the social order while transformation and social change take place" (1991, 75). The advent of social media has contributed to "the most important social change since the industrial revolution: the mediated communication order" (Altheide 2014, ix). *Process* is key. The emphasis on process helps distinguish and clarify social control from other generic definitions (as noted above, see Gibbs 1977). Process refers to "how something new and different [such as social media] becomes internalized and eventually assumed and taught as something everyone knows" (Altheide 1995, 6). Chapter 2 explores this process in relation to social media platforms and the impact of this technology on the public face of police and policing. The focus on institutional emphasis is significant as process "resides less in individuals than it does in processes and communicative logics that inform how situations are defined, the sources of such definitions, and their consequences" (Altheide 1995, xi; see also Altheide 2014). The concept of social control is also useful "as a *method* by which to study (or to interpret data about) social order" (Meier 1982, 35, emphasis original). As Altheide notes, "Social order is increasingly an electronically communicated and mediated order" (1995, xi).

SOCIAL CONTROL AND THE DEFINITION OF THE SITUATION

Social control involves the ability to define social situations (Altheide 2002; Goffman 1974, 1959; Thomas and Thomas 1928) and the capacity or social power to enforce a definition to encourage others to act accordingly (Altheide and Snow 1991; Altheide 1995). The definition of the situation then is an essential element of control. Despite the significance of the definition of the situation as a contribution to the sociological perspective (Bakker 2007), few have explored this approach since the 1970s (Altheide 2000). Definitions are important as they help frame social circumstances—namely, for our purposes, police legitimacy. No scholarship has contributed better to our conceptual understandings of how definitions frame social circumstances than that of Canadian-born sociologist Erving Goffman.

Goffman contends, "definitions of a situation are built up in accordance with principles of organization which govern events" (1974, 11). The progression of the Internet, text messages, digital selves and identity construction (see Zhao, Grasmuck and Martin 2008), and online relationships entirely removed from face-to-face contexts, for example, have rapidly advanced a type of social organization that Goffman and others who are concerned with social organization and control could not have foreseen. We are still learning how to make sense of developments in contemporary social control efforts, which now include how social cues are made available to a much larger *and* interactive audience on social media. These developments are transforming how contemporary social situations are defined and controlled by police and law enforcement agencies (Schneider 2015b) as well as how police and publics are repositioned in relation to one another (Schneider and Trottier 2012; Trottier 2012b).

Social control is anchored in assumptions about interaction (Janowitz 1991). Contemporary social control endeavors are intertwined with changes to fundamental social concepts such as friendship and privacy. What does it mean to be "friends" on Facebook with the New York Police Department (NYPD)? What are the implications for police surveillance (see Marx 1988) and privacy of individuals in this "relationship," both local and abroad (i.e., beyond police jurisdiction, in this instance, of the NYPD)? The proliferation of communication and information devices capable of linking to social media platforms contributes to identifiable changes in context, audience, and interaction. Meanings that emerge on social media can transform how authorities, such as the police, and citizens alike define situations and, in turn, how definitions influence interaction online. There may even be claims that contest authoritative definitions, including police definitions. This is, of course, not to suggest that police definitions have never been questioned (they most certainly have), but rather that definitions can now be routinely questioned and challenged by a much larger, interactive audience in publicly accessible online social spaces.

Social control is accomplished as a symbolically constituted and negotiated social process (Goffman 1959). This process now occurs in online spaces where the audience is both larger and interactive. A key feature of social control (as process) is the expressed participation of the controlled group—that is, how the public comes to accept police definitional claims as true, authoritative, accurate, and so on. This is of paramount importance for authorities because "the definition of the situation ultimately lies with audience response" (Altheide and Snow 1979, 19). In such situations, the audience response must be increasingly managed so that it remains conducive to police interests and social order (Ericson 1982; Manning 1978). Social media now plays a vital role in this process (Lee and McGovern 2014, 113–40). Previous scholarship has examined the ways that mass media formats have contributed to the definition of the situation (see Altheide and Snow 1979, 1991; Altheide 1995), and other research has advanced our understanding of police control of media formats such as television (Doyle 2003) and news media (Ericson, Baranek, and Chan 1989), but less work has been devoted to institutional police management of social media as a feature of social control.

MEDIA FORMATS AND POLICE PRACTICES

A 2014 *Washington Post* article called "Prince George's Police Plan to Live-Tweet Prostitution Sting" helps illustrate a shift in institutional police social control practices:

> In the coming week, Prince George's police say, vice officers will take to the streets to arrest people suspected of soliciting prostitutes. But in an unusual twist, they're planning to give the public an inside look at the sting operation—detailing it live on Twitter. Police said on their blog that it's a way to show how the department is "battling the oldest profession." Tweeting "suspect photos and information," they said, will serve as a warning and a deterrent. But the plan has provoked a backlash. Critics suggest that the tactic is no more helpful to protecting the public than watching an episode of "Cops." . . . Despite the criticism, the department isn't backing down, said chief spokeswoman Julie Parker. "It's not meant to be salacious," said Parker, who added that the effort is about transparency. "The community has a right to know, and we want to share it with them." (Bui 2014)

There are two fundamental distinctions worth noting that separate police use of Twitter from the television show *COPS*. First, unlike *COPS*, the use of Twitter encourages narrative participation from spectators, which means that spectators can become an important and sometimes an unpredictable part of the police narrative (for example, where police might not retain absolute control over the interpretation of the event). Second, ongoing police use of Twitter has zero financial costs, which was important when agencies such the TPS sought to expand its social media presence because, according to TPS Sergeant Tim Burrows, "there simply was no money available" (Burrows 2014).[4] When *COPS* aired in 1989, it cost "around $200,000 an episode" (Doyle 2003, 33). Police use of social media, as a presentational strategy, is less expensive, even with startup costs and some continuing costs, such as officer training. Police use of social media is performative and far more immediate than traditional media, including television programs like *COPS*; indeed, it can be an almost instantaneous form of communication with the public. But this is only a more recent development of police use of social media.

In his outstanding 2003 book *Arresting Images: Crime and Policing in Front of the Television Camera*, Aaron Doyle (2003) advances our understandings of the social (Meyrowitz 1985) and institutional influence (Ericson, Baranek, and Chan 1989) of television on policing and, in doing so, amends a few theoretical perspectives that deal with how media influence various social institutions. These perspectives use the cultural influence of media logic (Altheide and Snow 1979, 1991) as well as the influence of news media on criminal justice institutions, including police and courts (Ericson, Baranek, and Chan 1989). Doyle concludes,

> We need to move beyond a narrow understanding of media influence as simply affecting the attitudes, beliefs, and behaviour of individual audience members. . . . As we move forward, I hope the ways of thinking about TV I have suggested here may be extended in various ways to help us understand how media shape institutional life more generally, and how we must be wary that the most powerful players in any context tend to shape the forms this media influence takes. (2003, 145)

I take up Doyle's call to move beyond narrow understandings of media influence to examine how the evolution of media forms such as television, capable of "being an on-scene record of events" (Doyle 2003, 5), has expanded to include the now ubiquitous presence of communication and information devices capable of recording and documenting on-scene events (see Brown 2013, 2015).

SOCIAL MEDIA AND SOCIAL LIFE

More than twenty-eight million Canadians are wireless phone subscribers (CWTA 2015). Video recording capability is now a standard feature of nearly every single manufactured mobile phone. Storytelling is no longer reserved for news media journalists. Citizen-generated stories that circulate online might include the interpretation of video footage filmed on mobile devices. Such footage can range from extremely trivial and inconsequential matters of daily life (e.g., cat videos) to some extraordinary encounters with police officers (e.g., brutality). All of this (and much more) has now become Internet fodder, where videos are shared and broadcast to the world. These recorded documents remain available for continued, simultaneous, and repeated on-demand viewing by almost everyone everywhere.

Emergent scholarship has started to more seriously and systemically examine police and social media, including police use of social media platforms to communicate and interact with publics on Facebook (Lieberman, Koetzle, and Sakiyama 2013) and on Twitter (Procter et al. 2013a, 2013b; see also chapter 4 of this book). Recent research has also addressed developments in "cybercrime" (Chang 2013; Leukfeldt, Veenstra, and Stol 2013), while other work has explored the challenges of social media for police work. A special issue of the journal *Policing and Society* (2013) titled "Policing Cybercrime: Networked and Social Media Technologies and Challenges for Policing" is one example and focuses more on the nature and origin of digital messages than understanding the impact of media formats on police practices.

Social media provide new opportunities for examining representations of social meanings, on the one hand, and institutional relations, on the other. The case studies in this book draw from the social process and production of documents (e.g., posts on social media), as empirical accounts, that can collectively help us to better understand social change in relation to policing. This analysis moves away from various "big data" concerns associated with policing and social media (see Proctor et al. 2013) in lieu of a methodological approach that focuses on meanings, perspectives, and thematic emphases (Altheide and Schneider 2013).

My thesis that social media alter institutional police practices represents a contemporary application of *media logic* (Altheide and Snow 1979, 1991). Media logic is a theoretical approach that consists of an "analysis of social institutions-transformed-through-media" (Altheide and Snow 1979, 7). This way of understanding media represents an epistemological shift that moves away from the transmission model of mass communication (Lasswell 1948), and the decades-long debates over the *effects* that media may or may not have over individual behavior. This near century-old and ongoing debate over the effects of media has resulted in "thousands of studies" that collectively attempt "to assess the power of media to influence behavior" (Grossberg et al. 2006, 315). While interesting, these studies and scholarly debates have largely ignored the important influence of institutions (Ericson 1991) and the rules and logics of communication (Altheide 1995), with some exceptions (see Couch 1984; Carey 1989; Meyrowitz 1985).

THEORETICAL PERSPECTIVE

This book takes a symbolic interactionist perspective to its analysis of social media and policing. This perspective aims to understand the meaning of social activities in daily life (Blumer 1969). Social activities occur in numerous situational contexts and increasingly in online spaces, which can complicate social interaction (Johns, Chen, and Terlip 2014) by creating "*new* forms of action and interaction which have their own distinctive properties" (Thompson 2005, 32, emphasis original). A number of these social contexts are almost entirely mediated with little to no face-to-face interaction. From these spaces and online interactional contexts emerge new types of social forms for empirical analysis.

German sociologist Georg Simmel refers to "sociation" (*Vergesellschaftung*)—the "life of groups as units" (1950, 26)—as that "which synthesizes all human interests, contents, and processes into concrete units" (4). Interactive media as emergent social forms have become more prominent in daily life. How these developing social forms, including social interaction within these forms, contributes to changes in socially constructed reality is not fully understood. Previous work has theorized this process in relation to the cyberself (Robinson 2007) and modernity and politics (Thompson 1995, 2000) as well as developments in policing (Goldsmith 2010), but further work remains necessary to address the role of media logic throughout social life (Adolf and Wallner 2011).

This book develops and advances the assertion that "media are the dominant force to which other institutions conform," a basic principle of media logic (Altheide and Snow 1979, 15; Altheide and Snow 1991). I wish to show how the logic of social media contributes to recognizable changes in the ways that social media formats alter contemporary policing practices. The point is to illustrate how social media emerge from within a set of social and historical circumstances where social media become the *primary form* that contributes to recognizable institutional changes in public forms of policing. Building on media logic, this book focuses on the dominance of social media in relation to changes in policing, where police agencies increasingly tailor their content, messages, and emerging institutional strategies to conform to a new logic of social media (see Meijer and Thaens 2013), a different, albeit "mutually reinforcing," logic (Van Dijck and Poell 2013). The premise of Altheide and Snow's (1991) media logic is that information technologies and the format of these technologies influence how information is constituted, perceived, and interpreted in social interaction (Altheide and Snow 1979, 1988, 1991).

As a "very prominent social institution in almost every society" (Frey and Eitzen 1991, 503), sport is a good cross-cultural example to help demonstrate the significance of media logic. According to James Frey and Stanley Eitzen, "The nature of sport has been changed by the media with its emphasis on display" (1991, 510; see also Altheide and Snow 1979, 217–35 for a discussion of media and sports). The logic of media has altered how most major sporting contests are conducted—that is, how the material of sport is constituted, perceived, and interpreted. Nearly every professional sporting event has been altered from its original form of play by media presence. For example, sporting contests have been reorganized around commercial breaks to accommodate media forms like television and radio. This reorganization of sport to accommodate media allows for ratings metrics to take on an increased importance so that "once a sports entity has been displayed on television and received the financial support from television, the sports organization is forever changed" (Frey and Eitzen 1991, 510).

The presence of media have led to other changes to sport, including various amendments to rules, game regulations, and scorekeeping. These changes directly accommodate media representations that eventually become "taken as reality, overlooking reality that it is a staged event mediated by commentary" (Frey and Eitzen 1991, 510). The initial introduction of video reply in Major League Baseball (MLB) in 2008 and expansion of video replay in 2014 is a good example of how media has changed sport. Amending the MLB official rules to include video replay allows for the possibility of disputed calls during a live contest to be reversed. According to the MLB Replay Review Regulations, "If there is specific video that allows a Replay Official to definitively conclude that the call should be overturned or confirmed (as opposed to letting the call stand in the absence of video that provides clear and convincing evidence to overturn it), the definitive video used by the Replay Official in making his decision will be sent to the ballpark (to be accessible by the television broadcasters and scoreboard operator)" (2014). Media, as the above examples illustrate, are the principle force to which institutions such as sport conform. The specific example of video replay in the MLB influences how information about the game itself changes the nature of the game, including how the game is organized, played, and officiated. While sport and policing are distinct institutions, the premise of media logic remains, as each is reorganized according to select media principles.

SOCIAL MEDIA AS A SOCIAL FORM

Media logic views media as a *social form* that itself governs the possibilities of interaction. This perspective of media uses Simmel's approach (Altheide and Snow 1979, 1991), or *formal sociology* (Simmel 1950, 21–23), which avoids sweeping generalizations of the social world by focusing on an analysis of the "societal forms themselves" (such as media) as forms that "are conceived as constituting society" (Simmel 1950, 22). That is, formal sociology is concerned with how the logic of media transforms institutions such as politics or sports (Altheide and Snow 1979; Frey and Eitzen 1991). Formal sociology illuminates how a few of the recent institutional transformations in police and policing can be attributed to social media. Some research has examined the emergence of social media strategies in the public sector in government (Mergel 2010) and even policing (Meijer and Thaens 2013), but this corpus usually does not fully consider social media as a social form—one that alters police and policing practices.

The examination of media forms draws a conceptual distinction between the analysis of form and the study of content, or what Simmel himself calls the "material of sociation," where form is stable and content is variable (Simmel 1950, 40-41). "Simmel's image of sociation, however, is interaction with both form and content" (Duncan 1959, 101, cited in Lundby 2009, 109). Interactive social media provide new opportunities to empirically develop media logic in stronger relation to both form *and* content, the latter of which has been given less priority by Altheide and Snow as well as by Simmel himself (Lundby 2009). This book develops this area of inquiry by focusing on form *and* content and contributes more generally to developments in media sociology (Hjarvard 2013; Innis 1951; Lundby 2009; Meyrowitz 1985).

Much of the media research over the last several decades has focused almost exclusively on content or the "media effects" approach to the study of media, an approach that ignores the importance of form. A media effects approach focuses on the impact of media on the individual actor, and in doing so, fails to conceptualize media as a topic and resource of investigation. Neither does this line of reasoning necessarily account for both individual and structural relations because media effects scholarship ignores the importance of media relational forms, and thus illustrates the continued significance of Simmel's sociology of forms analysis (Simmel 1950).

A shortcoming of the media effects approach is that it also ignores the social implications that stem from a media logic approach—namely, identifying and examining the relational forms that arise from the dominance of media logic. This includes understanding how materials are organized to fit within the logic of a particular medium. In social media forms, examples include the 140 characters of a tweet, an image posted on Instagram, or a video on YouTube. Each medium is oriented to a different media format. Twitter developed from mobile phone Short Message Service (SMS) messaging and is primarily text-based; Instagram is an image medium; and YouTube a video-based medium. One can post text, images, and video on all three of these platforms, but how online users orient to each media format differs. In other words, users organize materials to fit each medium.

These orientations are connected with the use of grammar as form, as first noted by Simmel (1950, 22) and later developed in relation to media by Altheide and Snow (1979, 10). Altheide and Snow aimed to illustrate how media and its logic (as a relational form) underscore the construction of social reality. I locate and examine these consequences as they emerge in relation to the ways that the logic and form of social media formats has altered police work. The materials throughout this book demonstrate how the applications of the rules of media logic can apply to specific kinds of communication situations that involve police and recent innovations in information technologies.

MEDIA LOGIC AND DIGITAL MEDIA

A large movement of researchers, including, increasingly, those in communication studies, rather than, say, sociology, is now examining the basic principles of media logic. These advancements in the study of media, many of which have been recently developed by European scholars (e.g., Hepp 2013a, 2013b; Hepp and Krotz 2014; see also the Centre for Media, Communication, and Information Research at the University of Bremen in Germany), occur around the process by which the principles of media logic take place. The specific focus of these developments is on various applications of media logic through different academic traditions. These applications have been referred to by other names, including "mediatization" and "mediatized." These developments are firmly grounded in the creative research about new digital media formats. Media logic is the overriding principle of recent developments that are directed toward understanding connections between media formats and how institutions reshape themselves to fit media needs.

Networked media might differ with respect to how information is presented and transmitted, a process where media logic does not seem to neatly translate into the emergent logic of new media (Finnemann 2011). However, specific applications of the general principles of media logic remain in the process by which changes in communication formats invariably impact events, actions, and social interactions. Additionally, in some cases, these changes can produce new communication formats. This can include developments of media logic in relation to social media (Van Dijck and Poell 2013). Other research has examined applications of media logic relative to changes that coincide with media formats like cyberspace (Dahlgren 1996) and the Web (Sumiala and Minttu 2010). More recently, Jose van Dijck and Thomas Poell (2013) have explored this logic in relation to the principles of social media. Research has also considered whether or not new media might bring an end to media logic (see Schulz 2004). Knut Lundby suggests that "the advent of digital media, which have opened such wide opportunities for social networking and such varied options for multimodal expressions that the general concept of media logic . . . has come to an end, unless one makes very specific qualifications about the actual social interactions, constraints, formats, etc. involved in the mediatization process" (2009, 116).[5]

I suggest that a refined conceptualization of media logic—that is, how changes in communication formats can impact events, actions, and interaction, and produce new ones (vis-à-vis social media and the police)—can help provide some fruitful contributions to some recent European developments. For instance, this book links to some contemporary discussions of mediatization by "integrating 'media technology' into a theory of communicative practice or action" (Hepp 2011, 2). In doing so, the materials in this book connect the institutional parameters of media logic with empirical examples of "technological moments" (Hepp 2011), specifically those in relation to police work.

SOCIAL MEDIA LOGIC

Dutch scholars Van Dijck and Poell's 2013 article "Understanding Social Media Logic" provides some good and notably distinct applications (i.e., qualifications) of the principles of media logic to new media, although these particular qualifications do not appear to be offered in direct response to Lundby's above noted invitation to do so. Van Dijck and Poell refer to *social media logic* to discuss Internet-based applications that expand upon the basic foundations of Web 2.0, applications that allow the creation and exchange of user-generated content and have a set of "norms, strategies, mechanisms, and economies—underpinning its dynamics" (2013, 1). To elucidate the process of social media logic, Van Dijck and Poell (2013) identify four specific "grounding principles" of social media logic: *programmability, popularity, connectivity,* and *datafication.*

Programmability refers to the ability of social media site owners to alter specific algorithms in an effort to influence data traffic in what is often understood by users as an egalitarian and democratic medium. Popularity is "conditioned by both algorithmic and socio-economic components" and is governed simultaneously by both presence and influence (Van Dijck and Poell 2013, 7). Connectivity relies on networked platforms that connect users to advertisers. Lastly, datafication refers to the rendering of user activities, including use of and interaction on social media platforms, into data. These principles, Van Dijck and Poell argue, help "theorize a new constellation of power relations in which social practices are reshaped" (2013, 11). The authors, however, provide less specific qualifications about particular social interactions on social media, whereas the case studies in this book rely specifically upon these interactions. These patterns of social interaction (i.e., content) can be empirically located and examined to ascertain some of the ways in which social media (i.e., form) transforms relations between police and public.

Van Dijck and Poell's conceptualization of social media logic represents a good theoretical progression toward the development of a "succinctly different" form of media logic, albeit a "mutually reinforcing" (2013, 2) kind of logic. But this logic is not entirely applicable to an analysis of all social institutions, including the institution of policing. For instance, with social media forms, the balance of power shifts in some discernable ways *away* from police control—a basic theme of this book—whereas the power balance between oligopolistic media forms and police heavily favors various law enforcement control efforts (Doyle 2003; Chermak 1995). More specifically, one need only recall that police do not own the social media sites that they often use (e.g., Facebook, Twitter, You-Tube); nor do they directly control materials circulating on social media. They are thus unable to directly alter (e.g., programmability) specific algorithms on social media sites to meet their own objectives and interests. While police agencies can in fact manage posts to social media, management is different than ownership, as this excerpt from *Social Media and Tactical Considerations for Law Enforcement* illustrates:

> At one point, the TPS [Toronto Police Service] had to shut down the ability to post comments on the TPS's main Facebook wall, because the Service was unable to keep up with the large quantity of posts. Like many police agencies, TPS posts "Terms of Use" for its Facebook pages, stating that the TPS may remove viewer comments that are racist, defamatory, threatening, obscene, or otherwise "inappropriate or offensive." (COPS and Police Executive Research Forum 2013, 7)

An official statement on social media by the Royal Canadian Mounted Police (RCMP) further speaks to the point of limited ownership:

> The Royal Canadian Mounted Police's use of social media serves as an extension of its presence on the Web. Social media account(s) are public and are not hosted on Government of Canada servers. Users who choose to interact with us [RCMP] via social media should read the terms of service and privacy policies of these third-party service providers and those of any application you use to access them. The Royal Canadian Mounted Police uses: Facebook, Twitter, and YouTube. (RCMP 2014)

To further illustrate the point of the limited applicability of Van Dijck and Poell's social media logic to the policing institution, let us briefly discuss their concept of datafication. Police access to evidence via social media is greatly restricted by various international laws; while public interaction on social media sites might be rendered into usable data for police, legal access to private data remains limited. Social media, then, provides new opportunities and challenges for police agencies in Canada. Vancouver Police Department (VPD) Detective Mark Fenton, a recognized law enforcement expert on social media, explains a few of these challenges:

> Let's say you are in a bad relationship with your spouse. You break up and you are both on Facebook. Your ex-spouse is now creating a fake account on Facebook and is now harassing you on Facebook to the point where you are fearful for your life. So you come to us [VPD] and you say, "help me I want to prove that this is him, I know its him or her but I want you to prove it so we can go to court." Well, here is the problem. When you sign up to Facebook, did you know that all of your information is automatically assigned to be protected by the USA Patriot Act in the [United] States because Facebook is a U.S. company, so your data is protected by U.S. law? So even though we have a victim and a suspect that both reside in Vancouver, the information needed to

solve this crime is based in California. Currently, the law is that I have to write a warrant sworn here [in Vancouver]; [then the warrant] has to go to Ottawa if they decide it's ok it goes to Washington [DC]. Washington looks at it. If they like it, it goes to California, [and] if California likes it, then it goes down to the servers at Facebook. Then the reverse happens to make sure that the information is allowed to be sent to us [VPD]. It is not uncommon for this process to take *years* . . . and sometimes we don't even get it. (Fenton 2013, emphasis original).

Empirical research remains necessary to elucidate how social media logic contributes to the production of evidence and how these processes change the ways that law enforcement might secure warrants for access to restricted information on social media. More generally, however, my aim is to examine the transformative influence of social media in relation to sense-making strategies by police and public (such as detective work) and to then detail how police activities are altered by the process. As Van Dijck and Poell note, "social media logic *complements* mass media logic and enhances its dominant norms and tactics, just adding an extra dimension" (2013, 7, emphasis original). My conceptual use and treatment of social media logic is more consistent with Altheide and Snow's version of media logic—that information technologies, including social media, and the formats of these media forms govern how information is produced. In the chapters that follow, I wish to show that media logic has continued importance in the digital age. The next chapter explores how and when police become involved with social media. The chapter illustrates how police attraction to social media contributes to recognizable changes in institutional police practices that are consistent with the logic of social media.

NOTES

1. Nixle is a platform that allows authenticated government agencies, including police departments, to quickly send verified and secure messages to citizens. These messages are categorized in the form of alerts (urgent), advisories (less urgent), community information, and local traffic reports (www.nixle.com).

2. According to Dennis Brissett and Charles Edgley, "The most straightforward definition of dramaturgy is that it is the study of how human beings accomplish meaning in their lives" (1990, 2). This approach builds on Goffman (1959) and later Burke's (1965) insistence that the study of human behavior focus on action.

3. See Peter Manning's authoritative work on the application of dramaturgical theory on police work that elicits the articulation of "forms (such as etiquette, the legal code, traffic rules) that link routines, patterned actions, and structures" (1997, 8) in *Police Work: The Social Organization of Policing.*

4. While Twitter itself is a free service, costs associated with the TPS social media strategy included a $75,000 startup budget authorized by TPS Chief William Blair (Robertson 2014).

5. The term *mediatization* is popular in European-based media scholarship. Coinage of the term is attributed to Swedish media researcher Kent Asp (1986), who developed it from Norwegian sociologist Gudmund Hernes (1978) (Hjarvard 2013). Lundby

(2009), one of the most ardent critics of media logic, and others (see Lundby's 2009 edited volume) contend that mediatization is a distinctly different concept from media logic, even despite the fact that, as recently as 2013, "not much work has been done to define the term and develop it into a theoretical concept" (Hjarvard 2013, 8). Hjarvard, in his attempt to theoretically develop mediatization, notes "a contemporary parallel notion [of mediatization] in the work of Altheide and Snow" (2013, 9). Others have acknowledged that media logic has been reimagined and labeled "mediatization" to widen and expand developments that "lack a consistent definition of digital media" (Finnemann 2011, 67).

TWO

Crime and Society 2.0

Police and Social Networking

As media and communications studies researcher Daniel Trottier notes in his 2012 book *Social Media as Surveillance*, "social media features more and more prominently in policing and investigations" (138). This chapter traces when and how police in North America become involved with social media (see also Lee and McGovern 2014, 113–39) and explores how policing practices conform to the logic of social media.

Social problems, including crime, have developed, alongside contemporary forms of interaction online in recent years. We might refer to the relatively recent phenomenon of crime online as "crime 2.0." Understandings of crime have developed over time in specific historical and social conditions. What is crime in one place might not be crime in another. In other words, crime is situational and contextual. It can be "defined as a *social event*, involving many players, actors, and agencies" (Lanier and Henry 1998, 20, emphasis original). The advent and expansion of the Internet and social media bring together players and agencies where crimes can *simultaneously* come to the attention of everyone, across international and legal boundaries. The development of crime in online spaces can contribute to police decisions to invoke the law (see Goldstein 1960), especially in response to those activities that become "highly visible" in mediated spaces (Mawby 1999). This chapter traces the expansion of this process on social media. It is reasonable to assume that police activities would develop and expand on social media sites in direct response to concerns about crime, to match public expectations that "real police work is crime work" (Ericson 1982, 5). The question, then, becomes *how and when do police become involved with social media?*

WEB 2.0 AND CRIME

Time Magazine named "You" as 2006 person of the year. The "tool" that made this possible, according to the article, was the World Wide Web, more specifically the emergence and growth of Web 2.0. As with definitional concerns of "social media" discussed in the introduction, there remains some confusion and debate over what exactly constitutes "Web 2.0." The *Time Magazine* 2006 person of the year story avoids defining the term, opting instead to tell us, "Silicon Valley consultants call it Web 2.0, as if it were a new version of some old software. But it's really a revolution" (Grossman 2006).

Even while no concise or universally agreed upon definition exists, Web 2.0 can be differentiated from its predecessor Web 1.0 in a few ways. First, Web 1.0 is often distinguished in its capacity as a strict publishing platform, whereas a central hallmark of Web 2.0 is its user participation—hence "You" as person of the year. Users do not simply read a website as they did with Web 1.0; in Web. 2.0, they are also able to modify the content of the site. In other words, users contribute to and interact with the website. Web 2.0 is also distinguished by the idea of "the Web as platform"; that is, Web 2.0 does not have a boundary but a "gravitational core," which can be visualized as a "set of principles and practices" that help guide Internet business interests (e.g., search engine Google's algorithmic data management) (O'Reilly 2005). This characteristic was the initial guiding attraction of social media, which drew business interests, investments, growth, and later police and law enforcement agencies. Ian Eckert, a director of digital development, succinctly clarifies this perspective in relation to the relevance of sites like MySpace:

> Web 2.0 isn't about pushing out information to your audience—it's participatory and inclusive like MySpace or YouTube. The audience and the content is blurred and that's a real challenge. In the past, participation meant a letters page or a feedback slot, but that's essentially one-way traffic. Now, it's two-way traffic where people can gather around a brand 365 days a year. (Qtd. in Jones 2007)

A 2006 article in *MicroScope*, a UK computer industry magazine, noted, "A few sites grab attention and quickly show just what Web 2.0 is all about." On a short list following this statement, the author includes MySpace (Quickie 2007). MySpace (and its focus on music) helped spark and drive the popularity of Web 2.0. But the news was not all good. Along with Web 2.0 came crime, or "crime 2.0," what some have called "cybercrime" or "e-crime."

Crime is a feature of "all societies of all types," as Émile Durkheim (1982, 98) authoritatively asserted. If we accept this idea, it is thus logical (and normal!) that crime, including new crimes and responses to these crimes established by technological innovations such as Web 2.0, would emerge online. Our task is to understand how police agencies adapt and then incorporate these technologies into police work to deal with crime 2.0. This chapter focuses on this general process as it develops. The purpose is to examine how media logic is linked to developments in social media logic, a process that begins to materialize first with the growth of MySpace and develops on other sites, including Facebook, Twitter, and YouTube.

MYSPACE: "A PLACE FOR FRIENDS?" OR AN ONLINE HUB OF CRIMINAL ACTIVITY?

MySpace billed itself as a "place for friends," and, for a time, was one of the most popular sites online (once even more popular than the search engine Google). Its stratospheric popularity was relatively short lived, however. MySpace launched in 2003 and user popularity peaked between 2005 and 2007. While it was not the first social networking site per se, MySpace is important because it was the first to draw accelerated police attention to social networking sites.[1] MySpace was among the first of these sites to join music, blogging, and photographs, among other features, together within a single format. The unification of these formats, with music at its center, turned out to be the definitive ingredient of its early success and extreme popularity, especially among youth. A *New York Times* article characterized the site as "an online version of a teenager's bedroom" (Williams 2005). Because MP3 audio files could be easily uploaded to band profiles, MySpace provided a direct online venue to promote music and interact with fans, which effectively eliminated the pen-and-paper snail mail music fan clubs of yesteryear.

Each MySpace profile came with a unique web address to route users directly to pages that were customized according to the preferences of each individual user or band. In addition to a user name, either real or fictional, which was required to initiate a profile, each user could then modify their profile with HTML to include customized banners and graphics, images (without obscene content), videos, blogs, and, importantly, music, including individual songs that could be set to play automatically each time the profile was visited by another user—somewhat akin to the selection of mobile phone ringtones as a basic feature of youth identity management (see Schneider 2009). This user-driven interface made music discoverable, but more importantly, in terms of the popularity of MySpace among youth in a post Napster era, provided the means for users to copy, share, and, in many circumstances, legally acquire (i.e., download) music (MP3s).[2]

The site featured more than two hundred thousand band profiles by mid-2005, which attracted the attention of major media conglomerate News Corporation. A large percentage of MySpace users were teens and young adults—a notably elusive demographic for the purposes of selective marketing strategies. In July, News Corporation purchased Intermix Media (the parent company of MySpace) for $580 million (Kittler 2011). Reports of the acquisition brought further attention to the site. At the time of acquisition, MySpace had sixteen million users, and this number would nearly double by the end of 2005. News reports of notoriety associated with the site soon followed. For example, just weeks following the acquisition, twenty-seven employees were terminated from the Automobile Club of Southern California after just one worker complained of harassment in response to comments posted by these *off-duty* employees to their personal MySpace profiles (Associated Press 2005; Sprague 2007).

Much of MySpace's notoriety was linked directly with youth, the core user base of the site (boyd 2014). In the fall of 2005, news media reports of crimes (often *possible* crimes) (boyd 2014) in connection with teen use of the site began surfacing with more regularity. These reports largely focused on the site as an attraction for sexual predators. Research suggests that the modern moral panic surrounding the proliferation of sexual predators online was indeed exacerbated by MySpace (boyd 2014; Marwick 2008), and occurrences of sexual predation were actually infrequent. An April 2006 press release by the Federal Bureau of Investigation (FBI) stated, "despite all the media attention, criminal incidents are rare on these 200 social networking sites" (FBI 2006). Media attention directed in particular to MySpace had much to do with the format of the site, which enabled users to utilize search options to find music and locate friends.

For instance, one could search for other users of a particular age at a select school. This format, it was suggested across examined news media reports, provided online tools for stalkers and child predators to locate users (i.e., potential victims) by age (a minimum age of fourteen was required by MySpace to have a profile, however, reports highlighted that there was no way for the site to verify a user's actual age), music preference, gender, and location, among other search criteria. Predation was made that much easier because, in addition to searching for victims, predators might also learn a user's location or personal preferences, such as his or her taste in music, to facilitate the process of befriending ("grooming"), earning his or her trust, and arranging face-to-face illicit trysts. Concerns were likely exacerbated by the following "safety tip" featured on each MySpace profile: "That cute twenty-one-year-old guy may not be cute, may not be twenty-one, and may not be a guy!" News media repeatedly reported isolated incidents of teen crimes associated with MySpace, and along with accounts provided by claims-makers, often police officers, helped to normalize these activities. A *USA Today* report helps illustrate the point:

> Last month, for instance, a sixteen-year-old girl in Port Washington, NY, was molested after a man with whom she exchanged a few online

messages tracked her down because she had listed her workplace on her MySpace profile, says Port Washington Det. Sgt. Paul Gros. Police arrested a thirty-seven-year-old man, who he has been released on bail. But Gros says this incident, among others, should serve as a wake-up call. "As great a tool as (the Internet) is, there are a lot of risks that go along with it," he says. "Your freedom—and anonymity—isn't what people really believe it is when they use it. People may say it's not happening all the time. *Just be the victim of it once. Then it happened too many times.* (Kornblum 2005, emphasis mine)[3]

Other kinds of incidents involving teens were reported later in 2005. Many of these reports linked crimes associated with the teen use of MySpace but included adults as criminal co-conspirators. An *Observer* article draws our attention to this connection:

The madness of the internet and that of adolescent hormones can be an especially potent combination. Fourteen-year-old Kara Beth Borden of Lintz, Pennsylvania, met 18-year-old David Ludwig on MySpace. The relationship, which began as a friendship, intensified on the site. When her parents, who opposed the relationship, tried to cut off her online access to him, they were gunned down, allegedly by Ludwig, who then ran away with his underage sweetheart. They were arrested last month and it remains unclear whether or not Kara was involved in the killings. (Forrest 2005)

Reports such as these illustrate police attention to MySpace. This attention intensified near the end of 2005 in response to the high-profile murder of seventeen-year-old Taylor Marie Behl, a first-year student at Virginia Commonwealth University (VCU).[4] Behl was reported missing from campus by her dorm roommate at the beginning of the term in September. This is an early example where both news media and police simultaneously gleaned information from Behl's online profiles as the investigation into her whereabouts unfolded. Where police were once able to manage and (mostly) control information during investigations, including what information was released to news media, social networking sites made information freely available for use by others, including journalists. The availability of information on social media sites enables journalists to increasingly circumvent police as primary claims-makers (a process that is further explored in chapter 5). This practice has not only altered journalism but has also contributed to changes in police organizational procedure, such as how police perform contemporary investigations.

In the hours following the report of her disappearance, Behl's MySpace profile, referred to by police at the time as a "virtual tip machine" (Stockwell 2005), provided helpful information to aid police in their investigation. In fact, information from MySpace turned out to be more useful to police than the dozens of early tips they received. Determining these early tips inconsequential, police continued to direct their search efforts online for clues. As the *Washington Post* reported, "Now police also are privy to the disagreements that Behl had with her parents, her emotions on any given day, even her sexual exploits. By combing through the missing student's online journal and profiles, they learned not only about her favorite musicians and movies but also about the many people with whom she was acquainted on the Internet" (Stockwell 2005), including Ben Fawley, a thirty-eight-year-old photographer and former VCU student. Fawley was among Behl's ninety-two friends on her MySpace page. Police soon discovered that the two had been romantically involved. Following Fawley's arrest, which was initially unrelated to Behl's disappearance (although he remained a person of interest in the case and was later charged with murder), excerpts of their online exchanges were published in news media reports even while police refused to offer specific comment on the case because it was under investigation as a criminal matter. When photographs on Fawley's website reportedly led police to Behl's remains, online messages of sympathy quickly emerged on Behl's MySpace profile. Several of these user posts appeared in news media reports that discussed the case. Shortly after this case, the Richmond, Virginia, police offered a pilot program on youth Internet safety. Schools and police departments across the United States followed suit. Crime 2.0 had become lasting concern. There was no turning back.

THE EXPANSION OF CRIME 2.0 ON MYSPACE

Although MySpace launched in 2003, the site was not regularly associated with crime (or police) until later in the following year. A cursory search of the LexisNexis database between October 2004 and December 2005 indicates that in print news media featuring both "MySpace" and "crime," the latter word appeared just seventeen times; whereas, in reports with both "MySpace" and "police," the latter word appeared just fifty-two times. In 2005, the United States Justice Department reported that agents posing as underage users online had led to 600 arrests (Bahney 2006). Between 2006 and 2007, news reports that featured "MySpace" and "crime" and "police" ballooned astronomically to 518 and 815, respectively, increases of 2,947 percent and 1,467 percent. Over the course of the year, various news media reports that highlighted police officers posing on MySpace as underage teen users began to frequently appear.

Vigilante websites sprouted in the wake of these news reports. One site included the now defunct MyCrimeSpace where news reports of predators and related matters connected to child pornography on MySpace were posted. Vigilante groups, including those that even professed to work alongside law enforcement, also received growing attention. For example, perveted-justice.org, a federally registered charity organization founded in 2003, expanded its mandate in 2006 to "doing more and more stings with just law enforcement." This decision to expand their mandate, according to the Perverted Justice Foundation, resulted in hundreds of arrests and convictions of sexual predators (2015).

Concerns in news media reports intensified in February 2006 in relation to two Los Angeles-area teens who were reportedly abducted by a sexual predator using MySpace, allegations that turned out to be false (boyd 2014, 119–20). By April 2006, a *Newsweek* article reported that MySpace was responsible for 150 monthly criminal investigations (Romano 2006). A September 2006 *PC World* article called MySpace "a one stop shopping mall for online predators," naming the site as number one in a list of the top twenty-five worst websites online (Tynan 2006). In the ensuing moral panic, a *Dateline NBC* report put the figure of sexual predators online at any given moment at fifty thousand (Radford 2006). While this figure was not supported by evidence, it was nevertheless cited in numerous statements, including those made by some prominent claims-makers such as Alberto Gonzales, then United States attorney general. According to Benjamin Radford,

> News stories invariably exaggerate the true extent of sexual predation on the Internet; the magnitude of the danger to children, and the likelihood that sexual predators will strike. (As it turns out, Attorney General Gonzales had taken his 50,000 Web predator statistic not from any government study or report, but from NBC's *Dateline* TV show. *Dateline*, in turn, had broadcast the number several times without checking its accuracy. In an interview on NPR's *On the Media* program, [NBC Reporter Chris] Hansen admitted that he had no source for the statistic, and stated that "It was attributed to, you know, law enforcement, as an estimate, and it was talked about as sort of an extrapolated number.") (2006)

To address the growing concerns of sexual predators, MySpace announced in April 2006 that it had hired former Los Angeles County Deputy District Attorney Hemanshu Nigam to serve as their chief security consultant. Almost certainly by no coincidence, the appointment of Nigam was announced in the midst of a high-profile MySpace public service campaign launched in coordination with the Ad Council and the National Center for Missing and Exploited Children. Public service advertisements ran across News Corporation subsidiaries, including Fox. One of these advertisements reported, "One in five kids online is sexually solicited. Online predators know what they're doing. Do you?" (Newman 2006). This "alarming statistic," according to Radford, "is simply wrong" (2006). He continues,

> The "one in five statistic" can be traced back to a 2001 Department of Justice study issued by the National Center for Missing and Exploited Children ("The Youth Internet Safety Survey") that asked 1,501 American teens between 10 and 17 about their online experiences. . . .

When the study examined the type of Internet "solicitation" parents are most concerned about (e.g., someone who asked to meet the teen somewhere, called the teen on the telephone, or sent gifts), the number drops from "one in five" to just 3 percent. (2006)

Nonetheless, crime 2.0 flourished throughout 2006 when crimes associated with MySpace expanded to include an entire host of numerous illicit activities, such as cyberstalking, harassment, vigilantism, bullying, suicide and self-harm, identity theft, prostitution, pimping, computer hacking, drugs, defamation, child pornography, planned school shootings, extortion, terrorism, and various street gang activities, where self-proclaimed gang members with MySpace profiles were referred to as "Web bangers." One Florida Palm Beach County School District gang unit detective remarked, MySpace is "where I learned everything that I know about gangs" (Briscoe 2006). Also in 2006, Farah Damji, a sociology student at Open University, received the dubious distinction as England's first "on-the-run blogger," posting updates to her MySpace profile while on the run from a three-year prison sentence. Additionally, MySpace profiles of members of the Hells Angels, the notorious motorcycle gang, as well as profiles of death row inmates in Texas prisons were spotlighted in news media reports.

MySpace, a place for friends, had reportedly become a sprawling hub of criminal activity and attracted growing attention of the authorities, as did social networking sites more generally. While news reports continued to link social networking sites with criminal activities and untoward conduct, police responses to complaints remained largely reactive rather than proactive. Undercover efforts to apprehend sexual predators, for example, were more common than amending security policies on social networking sites. Police reaction to the chorus of complaints indicates that, despite growing concerns, the logic of social networking had not become fully integrated into police work. A 2006 *National Post* article pointed to the police's inability to address criminal complaints related to social media:

Ms. Aftab said her organization [Wired Safety, a volunteer watchdog group that monitors websites], which works closely with other social networking sites including MySpace and Facebook, has contacted the [social networking] site [Vampire Freaks] repeatedly and received countless complaints about its content, most from within Canada. "There are more Canadian teens using it than American teens," she said. "The major problem we're having with these sites is that it seems to normalize aberrant behavior." MySpace . . . as well as many other online communities, have drastically changed their policies to deal with security concerns. . . . *Staff Sergeant Gilles Deziel of the RCMP said they do not monitor the Internet generally, but might assist regional forces following up on specific complaints.* (Agrell 2006, emphasis mine)[5]

By the end of 2006, police use of social networking sites—MySpace, in particular—as a law enforcement tool had expanded, as did discussions about the use of social media within the law enforcement community. A November 2006 article in *Law Enforcement Technology*, a magazine for police, aptly titled "Catch a Creep: Come on Over to MySpace and You'll Solve Crime," provides a synopsis of crimes across the social networking landscape, with a few brief stories of selected criminal investigations. A notable shift in discourse occurs in the article: teens are portrayed as victims (of sexual predators), as is conventional, but they also increasingly become perpetrators of crime (e.g., drug use and weapons). The article notes that investigative data retrieved by police agencies from various social networking sites that does not necessarily fall within the parameters of crime might "involve a lesser offense, such as uploading inappropriate photographs," which is given as a reason for officers to "speak to the parents and show them what their child has been doing online" (Garrett 2006). Furthermore, despite reports of crime on social networking, "some agencies still resist employing these sites as investigative tools" (Garrett 2006). A few anecdotal accounts of "MySpace leads" are presented in the article as motivation for agencies to adopt social networking as a law enforcement tool.

The article also offers "tips [for police] to creating a plausible online identity":

- Showing vulnerability in the online persona. A weak or abusive father figure or parent-absent home shows a would-be predator the teen lacks a strong male figure or parental influence. To a predator this indicates the young person may be easily manipulated or exploited.
- Being isolated or lonely. Saying you're new to town, you can't make any friends, you're home schooled or suspended from school offers consistency—there's no reason to question why you're online on a Monday morning if you're home schooled or suspended.
- Exhibiting interest in alcohol or drugs. If a meeting is arranged, and the individual arrives with a case of beer for the teen, this shows intent and extra charges may be filed.
- Using age-appropriate words and grammar. "You have to sound like a fifteen-year old," [says Keith Durkin, a criminology professor at Ohio Northern University]. "It can't look like a third-year law student is writing it." The National Center for Missing and Exploited Children lists chat room acronyms to use. ICAC task forces also educate officers on pop culture. (Garrett 2006)

These suggestions, similar to news media reports that chronicled the rise of crime 2.0, speak largely to the efforts of *individual* officers and investigations, rather than specific police department protocols, policies, and developing procedures. Investigations also included challenges intro-

duced by the advent of social networking to foundational legal concepts
such as privacy (Petrashek 2010). However, many of these policies and
legal challenges would develop much later—in some cases, years later
after law enforcement agencies continued to realize and utilize the potential of social media. In an article on the website *Digital Trends*, Katie
Knibbs explains that

> Nancy Kolb, the program director who oversees the center for social
> media at the International Association of Chiefs of Police (IACP), says
> there were certain forward-thinking police departments that started
> venturing into social networks as early adopters during the MySpace
> days, but this really took hold later. "It was around 2009, 2010, 2011,
> where we really saw an exponential increase and huge growth in terms
> of law enforcement agencies using social media." (2013)

The "exponential increase" coincided with the growth of Facebook, now
the world's most popular social networking site. Facebook was launched
in 2004—one year after MySpace—but gained a more serious foothold in
the popular culture landscape in 2007, bumping rival MySpace to the
second most-visited social networking site. The exodus away from MyS-
pace was the very thing that had driven its early success: the door of the
online teenage bedroom had opened wide and parents and police were
now peering inside. This observation prompted Larry Magid and Anne
Collier, in their book *MySpace Unraveled: What It Is and How to Use It
Safely*, to note, "The bad news about this growing awareness of adult
scrutiny is that young MySpace users may just move on" (2007, 9). Recent
research shows that we are now witnessing a similar scenario with Facebook (Miller 2013). Whatever the reason for MySpace's decreased popularity, we would be mistaken to suggest that the site is no longer relevant
(as of January 2015, the site still had 50.6 million monthly active users
[Smith 2015]). What is important here, rather, is to recognize that it was
principally MySpace that attracted the widespread awareness of authorities to social networking sites, including Facebook.

THE RISE AND RAPID GROWTH OF FACEBOOK

Facebook, or "TheFacebook" as it was originally known, was modeled
after student paper directories that contained basic university student
information. Founder and former Harvard University undergraduate
Mark Zuckerberg launched an online version of the student directory at
Harvard in February 2004. The site was initially restricted to students
who were enrolled at Harvard, and only students with a Harvard–issued
email address could create a Facebook profile. Thousands of users at
Harvard registered and, by March 2004, the site expanded to include
other Ivy League universities, such as Yale. Soon after, Facebook allowed

university and high school students and members of companies to sign up (Locke 2007).

Initially, user profiles were arranged by university, school, and organizational affiliation. Largely for this reason, reports of negative or criminal conduct on the site usually involved or focused on various campus-related incidents, such as underage drinking or unauthorized parties. Between 2004 and 2007, except for these kinds of minor incidents, much of the negative attention in news reports—about crime and sexual predators, for example—was related to MySpace. This is not to suggest that Facebook avoided negative publicity as a niche university site. Criticisms associated with Facebook as "party oriented" remained a consistent theme across examined news media reports. This widespread belief pushed universities to adopt an institutional presence on Facebook for the purposes of student monitoring and surveillance (Monahan and Torres 2010). These surveillance tactics developed in reaction to "heavy drinking and wild parties [that] often accompany [the student] postsecondary experience," some of which circulated on student Facebook profiles (Trottier 2012b, 141).

The RCMP has referred to these gatherings (i.e., parties that involve youth that develop on social media) as "Project X" parties, so called after the 2012 Hollywood blockbuster movie of the same name, which a *New York Times* movie review described as "a teenage 'home alone' gone wild." According to an RMCP (Coquitlam, British Columbia) press release, these are "parties advertised through social media with the intent of attracting huge crowds of young people" (Chung 2012), despite the fact that the party depicted in *Project X* was not inspired by social media but by the online classified advertisements website Craigslist. Nevertheless, as Facebook expanded beyond use at universities, news media reports began to spotlight a larger range of criminal activities associated with the site, a development that appears to have coincided with increased non-university-related user traffic.

There has been much speculation about how Facebook succeeded at toppling MySpace from the perch of the social networking hierarchy. The topic, in fact, has inspired numerous bestselling books, such as *The Accidental Billionaires* (2009), and a Hollywood movie based on the book, 2010's *The Social Network*. Unlike MySpace, Facebook's growth was intentionally slowed by initial site restrictions. Furthermore, early on, Facebook was not guided by corporate interests; nor was it a priority to develop Facebook into a better business (Kirkpatrick 2010), whereas the opposite was true for rival site MySpace. Early Facebook users were not bombarded with corporate advertisements, which perhaps helped to attract an influx of users. Facebook was also not built upon a teen user base, a subtle but important distinction. Rather, the site began with a core user base of young adults (i.e., university students), a socially desirable group and one not believed to be disproportionately ripe with sexual predators and pedophiles. This difference in demographics allowed Facebook to avoid much of the early criticisms associated with other sites like MySpace; however, many of these negative criticisms emerged when restrictions on Facebook were lifted to allow teens to create user profiles.

The format of Facebook resembled existing social networking sites, but it was far more minimalist in design, as the site design moved away from customized banners and graphics like those associated with MySpace. This aesthetic difference presented Facebook as a seemingly more serious and hence "adult" alternative to other youth-oriented social media sites and, we might surmise, a more appropriate space for police to maintain official department profiles. The early intention of Facebook was to augment existent face-to-face relationships. This made Facebook stand out from other contemporary sites, including MySpace. As Facebook creator Mark Zuckerberg put it,

> Our whole theory is that people have real connections in the world. People communicate most naturally and effectively with their friends and the people around them. What we figured is that if we could model what those connections were, [we could] provide that information to a set of applications through which people want to share information, photos or videos or events. But that only works if those relationships are real. That's a really big difference between Facebook and a lot of other sites. We're not thinking about ourselves as a community—we're not trying to build a community—we're not trying to make new connections. (Qtd. in Locke 2007)

In September 2006, Facebook removed all its affiliation restrictions, abandoning the confirmed friendship model, and shifting to an open registration model. This move allowed anyone to join, including sensitive populations such as teenagers, but also the supposedly "fifty thousand" potential online sexual predators, at a time when MySpace was fully embroiled at the peak of a moral panic. The timing on lifting these restrictions and making an alternative and established niche site available to users could not have been better. Facebook quickly ballooned as users joined en masse. In one year, between December 2006 and 2007, Facebook grew from twelve million registered users to fifty million registered users. Facebook, once itself a niche site, had become so immensely popular that by 2010 it had pushed most rival social networking sites to develop a focus on niche user interests. Even the once mighty MySpace had opted to fully concentrate its efforts upon its original focus—music and entertainment, an established and strong core thematic of the site. At the end of 2010, the move prompted then MySpace CEO Mike Jones to firmly declare, "MySpace is not a social network anymore. It is now a social entertainment destination" (qtd. in Barnett 2011). By 2010, Facebook cemented its status as the dominant social networking site.[6]

Much like MySpace, Facebook had also attracted considerable attention from law enforcement for the purposes of investigations, but with an important difference and development: unlike on MySpace, police departments increasingly started launching Facebook profiles in larger numbers beginning in 2010. This development clearly signals the growing acceptance of the use of these sites among law enforcement. This is not to imply that police had not maintained a previous presence on sites like MySpace; some departments in fact had, but police activity, including interactions with the public, on these sites was minimal compared to current police activity on Facebook.

Police presence on MySpace and social networking more generally increased noticeably in 2006, at the peak of the fervor over sexual predators. This presence was largely for the purposes of criminal investigations. A confidential June 2006 *Law Enforcement Investigators Guide* issued by MySpace illustrates the point. The confidential guide, leaked in 2009 by whistleblower website WikiLeaks, was issued "for use by bona fide law enforcement agencies" and provides instructions to "serve law enforcement's needs" in an effort to assist with investigations, including the lawful acquisition of MySpace user data (MySpace.com 2006, 3). The guide brings to our attention what is perhaps today glaring obvious to many law enforcement agencies: that a vast amount of user data is publicly available. The guide's "General Information about MySpace and Law Enforcement Requests" section explains that

> MySpace receives a number of requests for information that is publicly available and can be obtained without the need for legal process or assistance from MySpace. MySpace profiles can be searched directly from the MySpace.com home page. MySpace also has extensive help pages that may assist law enforcement in determining if the information is publicly available, and may further assist law enforcement in understanding the particular features offered. (MySpace.com 2006, 4)

A similar set of online guidelines are publicly available on Facebook,[7] which can also be read as an indicator of the growing acceptance of the use social media materials by police as well as in legal proceedings. The MySpace guide is important because it placed the responsibility for knowing how to use online social media materials to aid investigations squarely on law enforcement agencies. The learning process that law enforcement needed to follow involved using MySpace in precisely the same way as the site's users did: sign up for an account, learn how to use the search function, initiate and accept friend requests, etc. At this time, few, if any, polices existed to govern the private, off-duty and public, on-duty police officer use of MySpace.

THE EXPANSION OF POLICE ON SOCIAL NETWORKING

Police use of MySpace beyond the purposes of police investigations began emerging in news reports in 2006. Much like everyday life transgressions of the status quo might draw police attention, the same might be suggested of posts made online by officers (on duty or off duty) that are not of a mundane nature. That is, these posts might violate shared social expectations of individual officers or may include information not usually shared with the public, or both. The 2006 arrest of country music star John Michael Montgomery illustrates the point. The arresting officer publicly discussed the arrest on his personal MySpace profile, which attracted scrutiny of posts made by four other officers, including disparaging posts about mentally disabled persons and homophobic remarks. These posts led to disciplinary actions and the termination of Officer Joshua Cromer (Brunty and Helenek 2015). The "Cromer Case" has been referred to as "the most widely known example of an officer coming to grief because of something that he or she wrote on a social networking site" (Scoville 2009).

Police use of social media to directly develop community relations came significantly later than the use of social media for investigations. The Los Angeles Police Department, the third largest municipal police force in the United States, for instance, launched a MySpace profile toward the end of 2006 to boost recruitment. Other police departments also launched MySpace profiles to establish a police presence on the site, but only later would become more active on social networking site Facebook. Amanda Fries outlines a timeline of social media adoption among smaller New York state police forces in a 2012 article in the *Utica Observer-Dispatch*, based in the Mohawk Valley, New York:

> Herkimer police Officer Jody Wheet said his department first delved in social media with a MySpace page about six years ago [in 2006] with the hope that children would add the department as a top friend, which hopefully would ward off online predators. About four years ago [in 2008], he said they started the Herkimer Police Facebook page, and it's mostly used to link to media stories, post safety tips, state crackdowns and surveillance videos. "As soon as we post (videos) on there, someone eventually will give us a tip or information that will lead us on where to go," he said. [Sergeant Steve] Hauck [Utica Police public information officer] said that Facebook most notably allows the department to have direct feedback from people. (Fries 2012)

This article helps to illustrate the shift from public *presence* (on MySpace) to increased *activity* (on Facebook). Here, police use of Facebook invites a more active response—the generation of crime tips—than MySpace, where police merely hoped users would add them as a "top friend." This development can be attributed to the coming together of information technologies, communication formats, and social activities (Altheide 1995). Although police agencies with official department profiles on Facebook are usually active on other social media sites such as Twitter, for the most part, they do not maintain similar or representative department profiles on MySpace (Lieberman, Koetzle, and Sakiyam 2013). At the time of this writing, Facebook is the social networking site of choice for police agencies. In 2011, the year following the rebranding of MySpace as a "social entertainment destination," the International Association of the Chiefs of Police (IACP), "the world's oldest and largest nonprofit membership organization of police executives," reported that more than 75 percent of police departments in the United States preferred Facebook.

Even as Facebook has remained the preferred social network for years, police departments have been much slower to join the site than other social institutions, such as universities (see Trottier 2012b). Police can be hesitant to join these sites because of the possible consequences of their visible online activities, which might include "risks to police integrity, effectiveness and reputational standing" (Goldsmith 2015). Despite these associated risks, many police agencies remain active on social networking sites, and public reaction to police social networking profiles has generally been quite favorable. The New York City Police Department (NYPD) Facebook profile is one example.

INCREASED POLICE ACTIVITY ON SOCIAL NETWORKING

The NYPD profile, according to the *New York Post*, went "live" on June 14, 2012,[8] and gained more than twenty-six thousand "likes" or endorsements in less than twenty-four hours (Auer 2012). Prior to the profile launch, however, social media had been the subject of NYPD investigations, and the department had a dedicated "social media unit." In August 2011, it was reported that the NYPD had formed this special unit "to mine Facebook and Twitter for mayhem" (Parascandola 2011). This unit, housed within the NYPD Juvenile Justice Division, was created "for the purpose of using social networking and intelligence to combat violent crime perpetrated by youthful offenders" (COPS and Police Executive Research Forum 2013). In response to the rapid expansion of communicative technologies across the spectrum of everyday life, law enforcement agencies worldwide continue to develop and incorporate similar units. However, this development does not tell us much about how social media contribute to changes in the organization of police work, including the contemporary use of communicative technologies in direct citizen encounters, an issue explored further in chapter 4.

Not all law enforcement agencies and police departments maintain active social networking profiles. Smaller police departments, for example, increasingly have a minimal presence or no presence at all on social media. Minimal or absent police activity on social media sites is usually the result of time management issues, including limited staff, or insufficient understanding of how to navigate these media, and quite usually both. Following the April 2010 York City, Pennsylvania police launch of Facebook, some of the department's officers suggested that the department should continue to expand its new social media activities to further direct citizen encounters. Police Chief Wes Kahley, who was updating the small-town York City Police Facebook profile himself from his living room, responded to the suggestion of starting a Twitter profile: "I don't know when I would find the time, keeping up with Facebook is hard enough" (Opilo 2010).

A contemporary characteristic of police is twenty-four-hour service; this key feature "distinguishes *modern* policing from earlier forms" (Williams 2007, 28, emphasis original). Modern policing, according to Richard Ericson, eventually gained acceptance following the "micro-level of everyday transactions with the citizenry" (1982, 3). It might be suggested that police use of social media, while still in its early stages of development, represents the most recent stage of micro-level interaction with citizenry, a form of interaction that helps maximize favorable impressions across a diverse audience (Manning 1978) and, importantly, over the course of a twenty-four-hour period. Forms of twenty-four-hour mediated impressions did in fact develop previously (e.g., CNN). However, with social media, police are now more fully able to actively control this process. On the other hand, police use of social media also includes new forms of the loss of control, as further explored in chapter 5.

Additionally, social media facilitate the necessarily "conveniently elastic" spread of crime control and now crime 2.0 control across the media landscape, a strategy that contributes to the reproduction of social order (Ericson 1982). The control of crime 2.0 involves the online process of acquiring and transferring information about policing as well as the management of public accountability (Ericson 1995; Mawby 1999). For law enforcement agencies, these concerns have led to various police departments hiring designated social media officers for the explicit task of maintaining department profiles, discussed further in chapter 4. However, because modern law enforcement use of social media remains in its early stages, various issues have thwarted police use of social media, including liability concerns. The Troy Police Department, also in Pennsylvania, for example, was ordered by the city council to stop using Facebook because of liability concerns, despite having received prior approval to use Facebook from the Troy mayor. A report from the local daily newspaper explained,

> When asked for comment about the reason for the discontinuance order, council president Jason Hodlofski said that a few council members were in a committee meeting recently and heard about the police de-

partment's Facebook page. "A few of us were unaware that we had one," he said. "So, we wanted to check into the liability issues of the page, and how that correlated to the borough and the council and those types of legalities, so we asked them to take it down so we could get those answers." (Hrin 2011)

The last post made by Troy Police to Facebook acknowledged compliance with the order but did not address the specific reason: "At the ordering of the Troy Borough Council the Troy Police Department has been instructed to discontinue using FACEBOOK in all official and unofficial aspects" (November 11, 2011). When asked for comment on the order Troy Police Chief Kyle Wisel responded in a statement,

> "I am optimistic the Troy Police Department will have use of Facebook once again. There are a couple of matters needing addressing before it can go into effect. Ultimately, I ask our citizens to be patient," he continued. "If approved [by Troy Council], I anticipate the department's use of Facebook will be an excellent tool bringing the community closer to its police department for the purpose of public safety and also a system to provide information to our citizens." (Hrin 2011)

At the time of this writing, more than two years following the above post, the Troy Police Facebook page remains inactive. The Troy Police, it seems, may have jumped the gun onto social media, and this appears to be the rule rather than the exception for many police departments that are currently active on social media. According to Murray Lee and Alyce McGovern, "many of the earliest examples of policing forays into the social media realm were developed with little awareness or consideration of the potential legal, operational and managerial implications of social media engagement" (2014, 115).

SOCIAL NETWORKING AND POLICY ISSUES

Evidence suggests that the lack of clarity around media policies creates risks for organizations and individuals who use social media (Lee and McGovern 2014). A May 2013 study by the U.S. Department of Justice in conjunction with the Police Executive Research Forum, a think tank for police chiefs, reported that just less than half (49 percent) of eight hundred law enforcement agencies surveyed in the United States had a social media policy, while the majority (88 percent) of agencies used social media (COPS and Police Executive Research Forum 2013). Police agencies outside of the United States have also launched profiles on social networking ahead of policy that governs both private, off-duty use and departmental use. One example includes the Vancouver Police Department (VPD), which is discussed in greater detail in chapters 3 and 4. The VPD launched their Twitter profile (@VancouverPD) on December 9, 2010. In an administrative report that followed the launch more than fourteen months later, the VPD officially acknowledged that "the VPD does not have policy to guide employees with their personal use of social media, nor are the investigative and operational uses of social media governed" (VPD 2012). Considering the significance of communication and information technologies, lack of or limited policies that govern officer use of social media is curious, as "policy governing the conduct of an officer on and off the job and general orders have been prominent features of policing since Rowan and Mayne assumed the joint commissionership of the London Metropolitan Police in 1829" (Manning 1997, 288). Policy generation in policing is often a response to an "immediate, localized crisis and illustrates the fundamental ambiguities in the craft" (Manning 1997, 289). The "craft" now includes emergent forms of direct citizen engagement on social media. In some circumstances, the generation of policy can follow officer gaffes made on social media, particularly those that attract unwanted or heightened negative attention to the policing institution. As an illustration, we turn our attention to the NYPD.

With more than thirty-five thousand sworn officers, the NYPD is the largest police force in the United States. The NYPD, like other police departments, had been using social media for investigation purposes and for police–citizen interaction ahead of any department policy on social media use. In September 2012, more than one year after the NYPD's launch of its social media unit and nearly three months after the NYPD Facebook page went live, NYPD Police Commissioner Raymond Kelly issued the first reported guidelines for use of social media in a five-page internal memo to officers. These guidelines were to aid with police investigations. In March 2013, the NYPD reportedly followed with a "three-page order" of "strict guidelines" that "had been in the works long before," or, more precisely, "a year and a half after officers posted insulting Facebook comments about the West Indian American Day Parade" (Goodman and Ruderman 2013).

Officers made disparaging posts to a publicly accessible Facebook group page titled "No More West Indian Day Detail." Sixty percent of the more than 150 people that posted to the public profile were identified as police officers (Glaberson 2011) and 20 of these were active NYPD officers (Kemp 2012). According a *New York Times* report, the Facebook profile commentary "ran 70 printed pages" long and contained, at best, "raw and rude conversations" and, at worst, offensive and racist comments. Officer posts included those that referred to people as "savages." One post read, "it's not racist if it's true" (Glaberson 2011). The *New York Times* article remarked,

> The page—though visible to any Facebook user before it vanished into the digital ether—appears to have drawn no public notice until an obscure criminal case in State Supreme Court in Brooklyn last month, the gun possession trial of an out-of-work Brooklyn food-service work-

er named Tyrone Johnson. His defense lawyers put many of the controversial remarks before the jury. But when that too seemed to draw little
notice outside the courthouse, the lawyers, Benjamin Moore and Paul
Lieberman of Brooklyn Defender Services, provided a digital copy of
the Facebook conversation to *The Times*, saying it raised broad questions about police attitudes. (Glaberson 2011)

The March 2013 NYPD guidelines, which were not made public, reportedly dealt with official (on-duty) use and personal (off-duty) use of social
media. With personal use of social media, the onus is upon officers to
scrutinize their individual posts to "defuse any lurking social media
landmines" that might reflect poorly upon the officer and thus the institution (Goodman and Ruderman 2013). The report spotlighted the lack of
NYPD policy on the off-duty use of social media. As another *New York
Times* article summarized, "For years, officers faced relatively few official
restrictions on social media, where many proudly posted photos of themselves in uniform and listed their job as 'N.Y.P.D.' Indeed, the Police
Department has lagged behind other jurisdictions in formalizing rules for
personal online behavior" (Goodman and Ruderman 2013). The NYPD
guidelines "broadly match those adopted by other big city departments
around the country," and include instructing police to refrain from posting images of other officers and not to post any non-public information
such as photographs of crime scenes (Goodman and Ruderman 2013).
Professionally, department approval is now required of all local com-
manders prior to official posts made on social media. The *New York Times*
notes that the new guidelines would only be enforced when alleged violations by way of "troublesome postings" are brought to attention of the
department.

To stay ahead of officer revelations of troublesome postings, other law
enforcement agencies review the profiles of prospective employees. The
Virginia State Police, for example, asks applicants to log into their social
media accounts in front of an investigator for review. As of January 2014,
this employer practice has been banned in eleven U.S. states, but as Virginia State Police spokesperson Corinne Geller puts it, "The social media
aspect is just one small part of a very complex [qualification] process that
takes place, and when you're entrusting an individual with the powers to
enforce the law, we are going to make sure that we have vetted this
person thoroughly" (Qtd. in Dame 2014). Such precautions are not necessarily surprising given that the off-duty use of social media has created
various problems for the institutional image and reputation of police
(Goldsmith 2015; see also chapter 5). Yet the growing awareness (chapter
3) and subsequent use (chapter 4) of these media are also seen as beneficial to the police institution. The key question here, however, is not
whether or not social media are good or bad for law enforcement; the
concern, rather, is to understand the relevance and logic of social media

and the recognizable changes in law enforcement organizational apparatus that social media has rendered.

SUMMARY

Police department policies signal the growing importance of social media, indicate the police acceptance of these media, and, importantly, illustrate the incorporation of the logic of social media into the police institution. As an indicator of the institutional incorporation of this logic we might consider social media sites that have emerged exclusively for active members of law enforcement. These sites are intended for professional use—for members to share techniques, strategies, and related expertise. They are not for sharing off-duty related materials.

One example includes BlueLine.[9] Intended for official use by law enforcement personnel, this free site launched at the end of 2013. The site uses a model similar to Facebook's now defunct confirmed friendship model—that is, the verification of user status as an active member of law enforcement is required to join BlueLine. The site also features a newsfeed, a "wall" for user posts, and "like" and "share" options for users—all modeled after Facebook. The significance here is not the existence of BlueLine, but rather that police now use social media as part of their institutional strategies. In other words, social media logic—the content, organization, and style of policing materials—is at play. Content, for example, is now "tailored to fit media format" (Altheide and Snow 1979, 23). A 2013 Reuters report illustrates how events "are *decontextualized* from their shared meanings by members and *recontextualized*" (Altheide and Snow 1979, 90, emphasis original) within a social media format:

> BlueLine users can create or join customized groups, with names like Gangs, Narcotics, New Technology or Sex Crimes—and then "crowdsource" colleagues for help with general aspects of investigations. . . . If a gangs investigator in one department comes across an unfamiliar tattoo on a suspected gang member, the cop can post it to a Gangs network, and someone from another department may help identify it as the sign of a new crew. Members can search for each other by name, geography, expertise and interest. Data analytics companies are developing BlueLine applications, which will let users create databases—of gang tattoos or graffiti tags, for instance—and analyze them. (Reuters 2013)

Previous work has examined how various social institutions such as politics, sport, and journalism have been transformed through media logic (Altheide and Snow 1979, 1991). Research has also examined this process in relation to changes in policing (Doyle 2003; Ericson, Baranek, and Chan 1987), but scholars have yet to examine this process as it relates to social media and policing. This chapter traced the progression of this process, beginning with police attraction to social media, and then illustrated how the logic of these media quickly worked its way into the organizational apparatus of law enforcement. The chapters that follow explore three Canadian case studies on social networking sites—Facebook, Twitter, and YouTube. To illustrate developing relations between social media and institutional police activity, these case studies reveal how select attempts to make sense of events, including crisis events, become quicker and more media focused.

NOTES

1. The first social networking site SixDegrees was launched in 1997 (boyd and Ellison 2007). Other early sites included LiveJournal (1999) and Friendster (2002). When MySpace launched, hundreds of social networks existed; however, few were regularly, if at all, associated with crime.

2. From 1999 to 2001, online peer-to-peer site Napster allowed users to distribute copyrighted music en masse and was shut down by legal injunction (see A&M Records, Inc. v. Napster, Inc., 239 F.3d 1004 [9th Cir., 2001]; see also Schneider 2007, 3129–34). Universal Music Group alleged in a 2006 lawsuit that MySpace encouraged users to share copyrighted music without permission.

3. Evidence continues to reveal that such occurrences are rare. The Crimes Against Children Research Center found "by examining police records and interviewing youth [that] when adults employ the internet in order to commit a sex crime involving a minor, it rarely takes that form" (boyd 2014, 113).

4. See Milivojevic and McGovern's discussion of the death of Jill Meagher and the role of social media in this process as a more contemporary example that explores how Facebook affects "the process of creating news and defining public agenda" (2014, 26).

5. Sergeant Gilles Deziel was the media relations officer at the RCMP headquarters in Ottawa. This statement was made a full two months after the first reported case in Canada that involved an alleged sexual predator and MySpace.

6. For readers interested in a general historical overview of Facebook, see Kirkpatrick's *The Facebook Effect* (2010). For those interested in a more scholarly perspective, see Trottier's *Social Media as Surveillance* (2012b, 33–60).

7. See "Information for Law Enforcement Authorities" at https://www.facebook.com/safety/groups/law/guidelines/.

8. According to Facebook, the NYPD joined February 27, 2012. See the NYPD Facebook page at https://www.facebook.com/NYPD.

9. See https://bluelineconnect.com/.

THREE

Facebook and
the 2011 Vancouver Riot[1]

This chapter focuses on the Facebook response to the 2011 Vancouver riot,[2] one of the first large-scale crime incidents in Canada where social media played an active and significant role in framing, defining, and making sense of the event both online and offline, largely beyond the control of police (Schneider 2015b, 2015c; Schneider and Trottier 2012). The documentation of the riot in Vancouver on social media differs somewhat from that of other recent events, such as the riots that took place in London, England, in 2011. Social media was initially believed to have played a much more important (and negative) role in the London riots (Baker 2012b). Later, however, experts determined that social media actually had much less of an impact relative to other media forms such as text messaging on mobile phones and television (Lewis et al. 2011).

British criminologist Tim Newburn argues that "the [August] riots of 2011 [in London, England,] represent the most significant civil disorder on the British mainland in at least a generation" (2015, 1). Commentators writing after the riots pointed to a number of factors believed to have contributed to the mayhem, including, most notably, social media (Newburn 2015). According to sociologist Stephanie Baker, "It appeared that England was experiencing a nascent social phenomenon, the origins of which, while resonating with traditional issues of moral decline and deprivation, were attributed to novel technological conditions of the twenty-first century: namely, new media" (2014, 112). *Reading the Riots*, a full-scale study of the August 2011 riots in London produced by a team of writers from the *Guardian* and the London School of Economics, determined that, "contrary to widespread speculation at the time, the social media sites Facebook and Twitter were not used in any significant way by rioters. In contrast, the free messaging service on BlackBerry phones—

known as 'BBM'—was used extensively to communicate, share information and plan in advance of the riots" (2011, 4). Social media played a more positive role. For instance, Twitter had an important role in the recovery following the England riots (Panagiotopoulos, Bigdeli, and Sams 2014). Riot Cleanup's Twitter page gained "more than 57,000 followers in ten hours" (Baker 2012b, 182). Very little evidence suggests that Twitter was used during the England riots to promote illegal activities (Tonkin, Pfeiffer, and Tourte 2012). In fact, empirical evidence suggests the opposite, that "the dynamic role of new social media [aided] in preventing, resisting and policing the crisis" (Baker 2012b, 169). The most important communication platform used during the 2011 England riots were BlackBerry mobile phones (Baker 2012a; Newburn 2015), and

> despite the attention paid to social media by government and the press, the *Reading the Riots* research suggests traditional media, particularly television, played a large part. More than 100 of the project's 270 interviewees referred to hearing about the riots via pictures on television news—more than Twitter, texts, Facebook or BBM. Some rioters also said the dramatic nature of the TV coverage tempted them to get involved in the unrest. (Lewis et al. 2011, 33)

In contrast, the citizen response to the June 2011 Vancouver riot on social media sites was unprecedented. Where *Reading the Riots* report clearly indicates that it was "BBM that actually played a substantive role in the [England] riots" (Lewis et al. 2011, 31), neither BBM nor BlackBerry appear even once across all three 2011 Vancouver riot reports. Moreover, response on social media during and after the Vancouver riot was more significant and had more of an impact than other media such as text messaging (SMS) or television. According to social media officer Constable Anne Longley, the citizen response on social media was "a response Vancouver Police didn't anticipate, *but had to adapt and respond to*" (Longley 2011, emphasis mine). The "response" to which Constable Longley refers is the volume of "users [that] drew on their social ties to identify and shame suspected rioters" on social media (Trottier 2012a, 416). As the external review of the riot (one of three official reviews of the 2011 Vancouver riot) noted, "As part of its operational planning process, the VPD did recognize they had failed to consider the potential role for social media during a riot situation. Clearly, this is an important consideration for police agencies across the country as the use of social media becomes increasingly mainstream" (Furlong and Keefe 2011, 291).

A few basic questions emerge from the response to the riot on social media: What can examining this process tell us about meaning making in relation to social events such as riots in online public spaces? And what might user activity on Facebook tell us about the social process of the definition of the situation, including the role of police in controlling the definition? Investigating some of the user responses on Facebook to provide insight into "how the events that night changed the way the department [VPD] uses social media" (Longley 2013), this chapter demonstrates that social media contribute to institutional changes in police work.

More specifically, this chapter explores how the most popular social media Facebook group page dedicated exclusively to the riot—the "Vancouver Riot Pics: Post Your Photos" Facebook page—contributed to the phenomena of collective interactive definitions (i.e., definitions of situations) that were beyond the full control of police following the 2011 riot in Vancouver, British Columbia. I argue that the unexpected response on social media served as a catalyst for police agencies, including the VPD, to develop institutional strategies that more fully incorporate the logic of social media.

THE VANCOUVER RIOTS

On June 15, 2011, the Vancouver Canucks hockey team was preparing to stave off elimination, in hopes of winning the National Hockey League (NHL) Stanley Cup, the most coveted prize in all of professional hockey, in what is arguably the most important sporting event in Canada. It was the final match in a best-of-seven series, winner take all. For many Canadians, when the local team is not in the playoffs, as a matter of national pride, you cheer for the Canadian team.

In Vancouver, the opposing team that evening was the Boston Bruins. The Bruins are one of the storied "Original Six" hockey teams of the NHL, making the contest all the more culturally significant. In Canada, 8.76 million people watched the game, which made it one of the most watched hockey games ever broadcast on the Canadian Broadcasting Corporation (CBC), second only to the 2002 Olympic hockey gold medal game between Canada and the United States.

Prior to the start of the June 15 hockey game, a reported one hundred thousand people "flooded downtown," with an additional "55,000 people in the fan zone," which was also located in downtown Vancouver (Furlong and Keefe 2011, 17). While attracting a crowd that size to the downtown corridor for the final game, coupled with the presence of alcohol, was cause for concern among politicians and police, similar large-scale gatherings occurred during the 2010 Winter Olympics in Vancouver without major reported incidents. But this time, immediately following the Canucks' loss to the Bruins, a melee ensued. Some people overturned cars and set civilian and police vehicles on fire. Storefronts were damaged and looted. Bystanders who tried to prevent looting were assaulted by rioters; there was a report of a fifteen-person assault on one individual who tried to prevent rioters from looting a popular retail establishment (Dhillon 2012a). Likely for these reasons, the legal response to the 2011 Vancouver riot was "highly punitive" (Arvanitidis 2013).

The online reaction to the riot was immediate (Longley 2013; VPD 2011) and multifaceted (Trottier 2012a). Social media users launched a public condemnation of the riot almost immediately after it erupted and while the riot was in progress (Schneider 2015a, 2015c). According to a *Vancouver Sun* article published on June 17, 2011, the "Vancouver Riot Pics: Post Your Photos" Facebook page was created within "10 minutes of game end" (Robinson et al. 2011). Claims-makers, including Vancouver Police Chief Jim Chu, referred to rioters as "anarchists and criminals who appeared to be the same people involved in the pre-Olympic demonstrations" (CBC 2011d). The sentiment that rioters had planned these events was echoed by others, including Vancouver's Mayor Gregor Robertson, who claimed that they "came with a plan for destruction in mind" (CBC 2011d).

In an October 2011 statement, Inspector Les Yeo, the leader of the riot investigation team, noted that he "expected 500–700 people to be charged" in relation to the June 2011 riot, but later indicated that he believed this number to be closer to "300 people" (Dhillon 2012a). By September 2012, the VPD had recommended a total of 275 charges against people "accused of participating in the riot, which left more than $3 million in damages" (Dhillon 2012a). The scope of the riot was vast, which had largely to do with the viral spread of information via social media. News media, including the *New York Times*, initially reported the event as "one of the country's worst episodes of rioting in recent decades" (Klein and Austen 2011). The lasting impact, however, was less substantial than anticipated (Mason 2012); injuries were relatively minimal—only a reported 160 people received riot-related injuries—and there were no deaths (Dhillon 2011). While reported damage was "significant to some businesses [it] was limited to a few million dollars" (Mason 2012). Indeed, these damage reports pale in comparison to the June 1994 riot in Vancouver, "which caused damage and costs to the city amounting to $800 million" and resulted in a reported "200 people [who] suffered minor injuries" (Howard 1995), including one severe injury after a man was shot in the head with a police rubber bullet.

On June 14, 1994, the Canucks played against the New York Rangers in a similar decisive Stanley Cup playoff game. Following the Canucks' loss in New York, a riot erupted in Vancouver. Authorities blamed the presence of television cameras and news media, asserting that the presence of media had a "provocative quality" (Howard 1995). Police used force to disperse between forty-five thousand and seventy thousand people from the streets of downtown Vancouver, and this use of force was documented by new media (Doyle 2003). Following the riot, police requested raw television footage and photographs to identify those responsible. However, "police attempts to obtain tape from local media after [the riot] became controversial when their first application to the Courts for access to photographs and video tape was refused" (Marfleet et al.

1994, 54–55). Some news media outlets, most notably the CBC, refused to hand over footage to avoid serving as "deputies for police" (Canadian Press 1994). Despite this reluctance, non-compliant news media were eventually forced by court order to surrender their materials, which included almost two dozen videotapes from the CBC and film and negatives from other television stations BCTV and CKVU and local print media from the *Vancouver Sun* and the *Province* (Schneider and Trottier 2012). As newspaper columnist Denny Boyd remarked, "An unfortunate by-product of the street riot is the testiness that has developed between the police department and the Vancouver news media over this matter of film evidence, evidence that police feel they have every right to demand as a part of their investigation" (qtd. in Cernetig 1994).

Vancouver police edited the confiscated footage to include rioters, and excluded the depictions of the police use of force (Doyle 2003). The footage of "100 images of suspects" was then set up in video kiosks around Vancouver, including in some businesses that were damaged during the riot (Thorbes 1994). Citizens were encouraged to identify suspects by entering information like the name, address, employer, and school of the suspect into the kiosk after viewing the footage (Doyle 2003; Thorbes 1994). CBC Television reported that "some shoppers say they feel uncomfortable with the idea of ratting on a suspect that they know, [but the] Vancouver Police are betting most shoppers feel the opposite" (Thorbes 1994). While these and other similar police efforts were deemed successful by authorities (British Columbia Police Commission 1994),[3] the *Globe and Mail* reported that "television viewers have become weary of the police videos, often obtained by subpoena and search warrant, that spotlight individual rioters" (Lee 1994). Following the 1994 riot, the police were able to frame and control the presentation and interpretation of media footage successfully (Doyle 2003).

Collectively, the riots in 1994 and 2011 resulted in dozens of arrests and millions of dollars in damages to the city of Vancouver (CBC 2011c). Canada is no stranger to hockey-related riots. In 1993, people rioted in Quebec after the Montreal Canadiens hockey team won the Stanley Cup final against the Los Angeles Kings. The point is that aside from the Vancouver riot in 2011, social media were not a part of these previous events.

Recent scholarly attention has focused increasingly on the proliferation of social media and related social changes (Mandiberg 2012; Trottier 2012a, 2012b), with much attention to Facebook (boyd 2014; Bazarova 2012; Trottier 2014). While this attention is indeed necessary, with few exceptions (e.g., boyd 2014), this work often overlooks the collective online meaning-making process in relation to "crisis-producing events" (Manning 1992). Citizen responses online to crisis events like the 2011 riot in Vancouver surprised police (Longley 2011, 2013). In this regard, very little scholarship in Canada has examined the online meaning-making process in relation to large-scale criminal events, even given the significance of the online response following the 2011 Vancouver riot as acknowledged by police (City of Vancouver 2011; Furlong and Keefe 2011; Longley 2011, 2013; VPD 2011).

THE VANCOUVER POLICE DEPARTMENT AND FACEBOOK

The Vancouver Police Department (VPD), the second largest police force in British Columbia, had a presence on Facebook years before the 2011 riot. In fact, the VPD was one of the first police departments in Canada to join Facebook, having initially activated their account on April 15, 2008. Nonetheless, their presence on Facebook before the riot was minimal.

The June 15, 2011 riot in downtown Vancouver was one of the very first of its scale, where user-generated documentation of the melee circulated on social media, especially Twitter and Facebook, almost in real time as the riot unfolded in the streets of downtown Vancouver, and well before official police control and interpretation of these data. A September 2011 *Globe and Mail* article described a few of the implications of social media in the aftermath of the riots:

> The online outrage after Vancouver's post-Stanley Cup riots turned into a summer project for Lee Timar. The unemployed IT professional devoted 12 hours a day through July to identifying and posting images of suspected rioters on the "Vancouver Riot Pics" Facebook page. Mr. Timar said he and a handful of others screened hundreds of hours of videos, looking for clear face shots and putting together collages of rioters. By August, they had identified about 300 people. They also turned the images over to police. (Elash 2011)

Responses to the 2011 riot mark the emergence of citizen *and* police surveillance on social media, where the citizen response was unprecedented (Schneider and Trottier 2012). The CBC's radio program *The Current* reported that the "Vancouver Riot Pics: Post Your Photos" page was "the largest Facebook group . . . devoted solely to posting pictures of the rioters" (CBC 2011b). Elsewhere, law enforcement agencies framed the Facebook group page as vigilantism, despite the fact that the VPD benefited from the online exchange of information and citizen documentation of the riot (Furlong and Keefe 2011; Trottier 2012a; VPD 2011). The U.S. Department of Justice publication *Social Media and Tactical Considerations for Law Enforcement* reported, "There also was a negative use of social media after the riots: to organize vigilante campaigns. In Vancouver, a Facebook group page called 'Vancouver Riot Pics: Post Your Photos' posted hundreds of pictures of suspected rioters to be identified by the public" (COPS and Police Executive Research Forum 2013, 31).

The Facebook page and response on social media did not go unnoticed by the VPD. In fact, the VPD incorporated the use of Facebook as an investigative tool *following* the riot. Also as explained in *Social Media and Tactical Considerations for Law Enforcement*, "In Vancouver, a Facebook ad campaign was developed to specifically target the demographic best able to assist investigators. Over 160,000 15- to 27-year-olds received ads on their Facebook profiles that would link them to the VPD riot website" (COPS and Police Executive Research Forum 2013, 30).

A *USA Today* report published on July 8, 2011, a few weeks after the riot, further illustrates the growing significance of social media in terms of the strategic modification of VPD practices as the department moves toward a more proactive use of social media sites such as Facebook and Twitter: "Since the creation of the Facebook page 'Vancouver Riot Pics: Post Your Photos,' almost 102,500 people have 'liked' the page, with hundreds of photos and videos contributed. *Even the Vancouver Police Department has joined in, using Twitter to send the same message*" (Spiegelman 2011, emphasis mine). In just the two weeks that followed the riot, 12,587 user postings were made to the main wall of the "Vancouver Riot Pics: Post Your Photos" Facebook page. The page was very popular, receiving more than 70,000 likes in just twenty-four hours, and reached 102,784 likes on June 29, 2011. User activity (i.e., posts) on the page decreased by 97.22 percent from June 29 to August 28, 2011 (Schneider and Trottier 2012; Altheide and Schneider 2013). Users posted commentary, images, and pictures, and named suspected rioters on the Facebook group page.

The citizen response of posting pictures and videos on social media can, in part, be attributed to the inability of the VPD to respond quickly to the riot on social media, even given the VPD's claim that they "used social media in a variety of ways during the riot" (VPD 2011, 75). While evidence suggests that the VPD used Twitter to disseminate "key messages," including "real-time updates on crowd congestion, traffic tie-ups, and public transit" (Furlong and Keefe 2011, 46), it is worth highlighting that a full twenty minutes of police inactivity passed on *all* social media platforms following the start of the riot (Longley 2013). Constable Longley confessed, "it was the first time we used Twitter during a large-scale special event" (2013). Why the VPD were less active on other platforms such as Facebook, despite the significant volume of user activity, remains less clear. We might surmise that, because Constable Longley was the only officer in charge of all VPD social media accounts at the time, she focused her attention on Twitter as a type of dispatch platform to disseminate up-to-date information from police, an issue discussed further in the next chapter.

We do know for certain that, by their own admission, the VPD were largely unprepared to use social media to interact directly with citizens during the riot. Many months later, Constable Longley remarked, "now we're more prepared" (qtd. in Shaw 2012). She also reflected, "what I didn't know back then [before the riot] was just how important having a voice on social media is for police agencies" (Longley 2013). A post made by the VPD on their Facebook page just weeks before the riot, presumably by Constable Longley, provides one likely explanation for the minimal activity of the VPD on Facebook leading up to the riot:

> The VPD Facebook site will be a little "quiet" over the next few weeks as I will be away on vacation—exactly during the Stanley Cup Playoffs! During the playoffs our message to all the Canucks' hockey fans is to "celebrate safely and get home safely." Many officers will be working to ensure everyone has a safe and memorable time and will be "dusting off their high fives" along with you. Go Canucks! (May 26, 2011, 1:48 p.m.)

Use of social media to issue such a statement to the public—"celebrate safely and get home safely"—appears to be in line with police press releases published by traditional media outlets (Ericson, Baranek, and Chan 1989). One user response to this post read, "That was a smart time to take a vacation! Now you'll miss either drunken reveling *or a* riot" (May 26, 2011, 4:02 p.m., emphasis mine). It is worth mentioning that the VPD post was made on Facebook in the context of heightened discussions in national and local news media, just weeks prior to the 2011 riot, about the possibility of hockey riot similar to the one in 1994. The VPD Facebook page remained mostly silent during this time but, for reasons unknown, appeared active again in the days just before the riot. On June 13, 14, and 15 (before the riot), three apparently random and innocuous images of crowds were posted by the VPD.

Following the riot, seven unique posts were made on the VPD Facebook page by the VPD, including the following post that solicited documentation from the public:

> If you took photos or video last night of the riot suspects of people committing criminal acts, please follow the instructions in the link on where to submit them to the Investigative team. We are looking for your original footage, not footage you've found posted elsewhere on the internet, as we already have those links and investigators will be looking through them. Thank you! (June 16, 2011, 3:11 p.m.)

With little explanation offered by the VPD, the onus was placed upon citizens to discern what constituted criminal acts from non-criminal acts. Debates on Facebook ensued over the nature of "relevant" versus otherwise "non-relevant" user-generated evidence (Schneider 2015a).

Another explanation for the VPD's limited use of Facebook at the time was that they were too overwhelmed by the online response. The VPD later acknowledged as much in their internal review of the riot:

> Almost immediately people were requesting information on how to send this information to police through the Twitter and Facebook feeds. . . . The VPD [received] an unexpected and unprecedented number of people submitting evidence over social media. *The VPD asked people to hold onto their photos and videos because the VPD could not manage*

the information during the riot. . . . Many of those with photos and videos began making their own websites, blogs, Twitter feeds and Facebook pages to identify the rioters. (VPD 2011, 75, emphasis mine)

The riot did bring attention to the VPD Facebook page. Ten days following the riot, the VPD posted, "We've just hit 5,000 Facebook friends today—thank you all for being here and supporting the Vancouver Police Department!" Despite the noted significance of social media during the riot, as acknowledged in statements provided by the VPD as well as in the external riot report (Furlong and Keefe 2011), the VPD riot review concerning police use of social media actually says very little. Just four short paragraphs cover "The Impact of Social Media" in the hundred-page VPD riot review (VPD 2011, 75–76). This is surprising given the VPD's acknowledgment that "the huge number of photos and videos" provided by citizens were "certainly beneficial to police in their investigation" (VPD 2011, 75).

RIOTS AND THE DEFINITION OF THE SITUATION

Riots are criminal events that involve crowds, violence, and police, but the definitive criteria remain vague. The number of people necessary to constitute a riot varies, as does the amount or kind of violence displayed. All riots share two characteristics: they are defined as such, and they most always involve police or other state agents, such as military. How riots are understood and subsequently defined is of paramount importance because these definitions contribute directly to how these events are controlled by police and handled conceptually by state officials. Social media can complicate the justice process for authorities. As the *VPD 2011 Stanley Cup Riot Review* explained,

> The 1994 riot review noted the impact of TV news cameras on the behaviour of the people in the crowd. The 2011 riot can be distinguished as perhaps the first North American social media sports riot and the acting out for the cameras seen in the 1994 riot was multiplied many times more in the 2011 riot by the thousands of people cheering the rioters on and recording the riot with handheld cameras and phones. . . . The shaming of people increased to the point that the VPD felt the need to issue a statement to remind people to be patient and resist the temptation to take justice into their own hands.[4] (7, 76)

CROWD MANAGEMENT AND "SETTING THE TONE" ON SOCIAL MEDIA

Vancouver has seen two hockey riots: one in 1994 and the other in 2011, as we already know. Following the 1994 riot, the police maintained total control over the recorded documentation of the riot (Doyle 2003), but, in

2011, the police had almost no control over the circulation and interpretation of the documentation of the riot on social media (Schneider and Trottier 2012). In the words of VPD social media officer Constable Longley,

> Game 7 was the final and deciding game of the Stanley Cup, and the Vancouver Canucks had not been in that position since 1994, when they lost in Game 7 to the New York Rangers. In 1994, the hockey fans rioted, and 2011 turned out to be no different. As I sat in the Command Room, watching cars being torched and people destroying the buildings and streets downtown, I just stared at my computer and thought, "What do I say [on social media] now?" There was no template of what to do or what to say, and any of our prepared messaging was no longer valid. (Do I use a hashtag? Which one? Should I use the word *riot*?) (2013)

The circumstances of each riot might have been similar, but the response was far from the same. The 2011 riot is more significant than the 1994 riot in that the 2011 event increased police awareness in Vancouver, and across Canada, of the necessity to adjust institutional strategies to include the use of social media in response to the social upheaval in Vancouver. The 2011 riot "was a new phenomenon for the VPD," Constable Longley noted. "Social media was all of a sudden front-and-centre, with 'Identify the Rioter' Facebook pages popping up and photos being shared over and over again" (Longley 2013).

In the wake of the riot, the VPD's *proactive* use of social media to increase police officer visibility and interaction with community members grew. The VPD, for instance, "now use social media for every big special event in Vancouver" as a strategy to help set the tone of the event on social media (Longley 2013). In the external 2011 riot review, undertaken by the Government of British Columbia, authors John Furlong and Douglas J. Keefe note, "VPD's crowd management strategy is good but is predicated on setting and maintaining a tone of responsible celebration. The opportunity to set a tone passed before there were enough officers to set it and congestion prevented its later imposition" (2011, 3). Police efforts to set the tone beyond, say, a single post on Facebook that encourages fans to celebrate safely, have expanded beyond special events to the everyday. A policy statement from the VPD's *2012–2016 Strategic Plan* helps illustrate the point:

> The VPD has launched initiatives to improve communication to the public through the Crime Alerts Program, Twitter, and Facebook. These initiatives have allowed the VDP to communicate directly with the public, providing more information than ever before. Through social media, members of the public are also able to find out more about the VPD and how officers do their jobs on a daily basis. (VPD 2012, 25)

Furthermore, "teaching new [police] recruits about social media is now part of their curriculum when they start with the [Vancouver Police] Department" (Longley 2013). These efforts serve as presentational strategies that help facilitate police use of social media, strategies that are explored in more detail in the following chapter.

COLLECTIVE BEHAVIOR AND THE
DEFINITION OF THE SITUATION

Riots involve crowds (Wright 1978). According to Neil Smelser, "collective behavior is guided by various kinds of beliefs [including] assessments of the situation" (1962, 8). Collective assessments of situations like riots involve crowds. A crowd is understood to involve a gathering of people who share common values (Smelser 1962). Social activities, such as those that involve crowds, can now come together in mediated communication environments (Baker 2011; Schneider and Trottier 2012). In *Understanding Digital Culture*, Vincent Miller suggests that social activities in digital networks serve "as a more accurate depiction of relationships within late-modern society" (2011, 184). And while people use social media for a great variety of reasons, recent evidence suggests that, in response to contemporary "crisis-producing events" (Manning 1992) such as riots,[5] individuals mobilize in online spaces to interact with those who share their values. This type of collective action, emerging from social media and occurring on mediated platforms in response to riots, has been referred to as a "mediated crowd" (Baker 2012a). This phenomenon "traverses both physical and digital modalities" where "media do not simply represent reality, but constitute it" (Baker 2014, 112), which is consistent with the principles of media logic. There is not necessarily a shared universal consensus among users in a mediated crowd; rather, in such circumstances, "a *crisis* mode of communicating [about police on social media] draws community support" (Manning 1992, 146, emphasis original). This type of community support is evidenced by unprecedented responses on social media sites created and dedicated to coincide with "crisis events" in reaction to recent riots. For example, after the 2011 London riots, "more than 12,000 individuals were identified as mobilising support [via Twitter] for the riot clean-up" (Lewis et al. 2011, 32). Such responses on social media can be viewed as a digital version of Herbert Blumer's (1951) "acting crowd," a group that directs actions toward achieving a stated goal.

Online sites modify spatial and temporal elements of the acting crowd, including how users respond "directly to the remarks and actions of others" (Blumer 1939, 180). Such online sites operate as virtual spaces where social actors can make sense of and respond to social events, a dynamic interplay between the individual and the collective. What makes this relevant to the researcher of everyday life is how recorded documentation of this information serves as an accumulation of "definitions of situations," whereby the "attitudes of others are in fact the most important limitations upon the behavior of any one person, and they are also the most important inducements to any particular action" (Waller 1970, 163). The definition of the situation has been articulated as both the result and cause of collective behavior (Turner and Killian 1987). This scholarship, however, does not take into account the online processes through which social cues may contribute to definitions of situations, which in turn may contribute to online crowd behavior in response to select crisis-producing events (Manning 1992).

Understanding crowd behavior has been an area of interest for sociologists for over a century. Early development in this area can be attributed to Robert Park (McPhail 1989), who, along with his colleague Ernest Burgess, defined sociology as the "science of collective behavior" (Park and Burgess cited in McPhail 1989, 405). A student of Park, Blumer developed the study of "collective behavior" (1939). The analysis of collective behavior, for Blumer, was to advance understandings about "the way in which a new social order arises" (1939, 223). He asserted that human group life should be studied in the context of human action—more precisely, social interaction (1969).

Increasingly, much of our social interaction is occurring online in mediated spaces (Miller 2011).[6] Yet very little scholarship addresses online interaction with respect to the definition of the situation, which is an element of collective group behavior. As a matter of course, social interaction (online or offline) involves the social process through which humans transform themselves in situ. Two basic assumptions underlie this perspective: (1) people act jointly in groups, and (2) people act independently of one another while drawing upon a social framework of everyday expectations. Both of these processes contribute to collective behavior.

A plethora of scholarly work has examined and theorized collective behavior (e.g., Blumer 1939; Killian 1984; McPhail 1991, 2006; Miller 2000; Neal 1993; Park 1927; Smelser 1962; Turner and Killian 1957, 1987; Turner 1994, to highlight a few). The task here is not to provide a robust literature review of these materials.[7] Instead, I want to emphasize that the vast body of literature on the subject does not account for the expansion and increasing use of social media, and that examining collective behavior on social media provides some new empirical possibilities to expand this work, especially in relation to how collective behavior online might contribute to developments in institutional police strategies. This can include police agencies using social media to create "citizen investigators," but, according to Billy Grogan, chief of police for the Dunwoody Police Department in Georgia, "that will never happen unless your department is using social media and making a concerted effort to engage your community" (2014).

Investigating collective behavior can be approached via a diversity of theoretical "strands" (Marx and Wood 1975), which should include further empirical consideration of the definition of the situation as it develops in the "mediated crowd" (Baker 2012a). The definition of the situation is a perspective initially advanced by W. I. Thomas. According to Willard Waller, the definition of the situation "is the process in which the individual explores the behavior possibilities of a situation, marking out particularly the limitations which the situation imposes upon his behavior, within the final result that the individual forms an attitude toward the situation, or, more exactly, in the situation" (1970, 162). It is important to understand how people make sense of the possibilities for action (framed by norms), the limitations (possible deviations from those norms), and how people formulate a normative (or deviant) attitude. Social cues are a consequential characteristic of this process. A less discussed feature of the definition of the situation concerns investigating the manner in which social cues are provided to people. Robert Perinbanayagam (1974) notes that while W. I. Thomas indicates the necessity of social cues in the determination of the definition of the situation, he fails to explain how cues are made available to actors. Interaction through social media contributes to this process. A shared consensus about the definition of the situation can emerge through group interaction on social media. Waller suggests, "we may refer to these group products as definitions of situations" (1970, 162). Social media, of course, do not determine collective action; rather, social media provide a powerful social platform for individual statements given in the context of the stated attitudes of others "to create a common identity and mobilise collective action," as the use of social media during the 2011 England riots revealed (Baker 2012b, 186). The response to the 2011 Vancouver riot on social media is a case study through which we can understand this online social process in a Canadian context.

RIOTS AND THE ROLE OF MEDIA

A riot remains a legal classification that denotes some level of disturbance in the context of a collective gathering. The *Criminal Code of Canada* defines an unlawful assembly (i.e., "riot") as three or more persons, but the number of people necessary to name a gathering as a riot might differ elsewhere. In the United States, for instance, the state of California Penal Code defines an unlawful assembly as two or more persons. What exactly determines a riot and how subsequent definitions are constituted remains a symbolic process, and social media have contributed to recent changes in this process (Baker 2011). In "Policing the Riots: New Social

Media as Recruitment, Resistance, and Surveillance," Stephanie Baker writes, "New social media itself contributes to these interactive relationships by broadening the public space for communication and self-constitution with the medium perceived to be an 'extension of ourselves'" (2012b, 187).[8]

The negotiation of meanings in these situations is paramount, particularly for social control agencies such as police (Trottier 2012a, 2015). Riots involve acts of collective behavior; they are temporary social events. They can occur in many different symbolic contexts. Examples might include sports riots, race riots, and now a "new form of rioting" on social media that is distinguished by "the speed and scale . . . from previous incidents of civil disorder" (Baker 2012b, 170). The literature on race riots (see e.g., Bergenson and Herman 1998) has examined the past and present in terms of cultural-collective behavior (Turner 1994), various background causes (Lieberson and Silverman 1965), as well as the aftermath of these riots (Gordon 1983). Other work on riots has explored the dynamic between the individual and the collective group, which together consist of the formation of temporary gatherings (McPhail 2006). The role of media in the dissemination of images and accounts pertaining to riots, and the assignment of meanings, remains an important area of continued exploration, especially in relation to the role of social media (Baker 2011). Previous scholarship in the area of mediated accounts has investigated how riots are negotiated after they have occurred, and the importance of television media in this process (Doyle 2003).

While conventional media are of course not a determining factor of interpretations (see Ericson 1991), such accounts nevertheless contribute a great deal to the "retrospective reinterpretation" of riots (Doyle 2003, 87; see also Goldsmith 2010). Social media use provides an interactive platform for the circulation of accounts following riots that give rise to the "mediated crowd" phenomenon (Baker 2011, 2012a, 2012b). Prior to recent developments on social media, other media forms such as television or print have helped contribute to how situations are defined and understood across geographical boundaries.

Televised news coverage consists of aural and visual accounts. Together, these accounts contribute to the discourse surrounding riots, which, in turn, aids in the social construction of these events (Doyle 2003). "By seeking to disseminate information that people want, need, and should know, news organizations both circulate and shape knowledge" (Tuchman 1978, 2). In other words, news media are a feature of the meaning-making process, as the "mass media do not merely report on events but rather participate directly in processes by which events are constituted and exist in the world" (Ericson 1991, 219). The construction of imagery and definition, for instance, contributes to the "societal reaction" (Lemert 1951) that influences how these events are discussed in media (Young 1986). Media also bring awareness to criminal events and, in doing so, establish a "reality" of the event. In *Manufacturing the News*, Mark Fishman explains, "Public events have never been known apart from the institutionalized means of communication which formulate those events in society. In our age, these institutionalized means of communication are the mass media. They set the conditions for our experience of the world outside the spheres of interaction within which we live" (1980, 12). Let us consider an example of print media to help illustrate the point.

During coverage of the Haymarket affair, also known as the Haymarket riot, which happened in the aftermath of a bombing that took place during a labor demonstration in Chicago on May 4, 1886, the *Chicago Tribune* featured print reports that included selected circumstances of the occasion. Journalists report the "facts" of the event, and because of the structure and format of the print medium (i.e., limited text space), the coverage is, like with any medium, abbreviated. The coverage of the Haymarket affair, however, short of editorials, largely excluded the collective perspective, and provided little insight into the process of public understanding. Television coverage, especially live coverage, like print, is also abbreviated, which can also be attributed to the structure and format of the medium. Additionally, televised broadcasts of events and their participants are, also like print media, usually not interactive in that televised representations often do not take into direct account the interpretations of those involved in the event. There are, of course, exceptions. Televised interviews, for instance, might include reports from eyewitnesses, but rarely will one ever see televised interviews with those actually participating in riots.

Social media have added a participatory dimension to media coverage that gives rise to the emergence of "citizen journalism" (Brown 2013, 2015; Greer and McLaughlin 2010). The construction and distortion of the public event is no longer entirely the work of news journalists (Altheide 1976; Fishman 1980; Tuchman 1978) or police (Ericson, Baranek, and Chan 1989; Doyle 2003). Social media offer spaces for group participation in the meaning-making process alongside official definitions offered by police and state agents (Baker 2011, 2012a, 2012b; Trottier 2012a, 2015). Before these media, traditional mass media served primarily an oligopolistic function. Consumers were unable to interact directly with disseminated materials or with these media. Social media, such as Facebook, now operate as a platform for these forms of interaction to occur. The construction of the "reality" of the event becomes visible in these online spaces where accounts, excuses, and justifications (Scott and Lyman 1968) as well as reaction and commentary can be located, collected, and analyzed to better ascertain the meaning-making process (see Schneider 2015a).

Social media provide a platform that allows a spectrum of users, from agents of social control to the general public, to generate content and provide accounts, adding a new empirical development in the examination of the "mediated crowd" (Baker 2012a). These developments can include expanding our understandings of collective behavior and the roles of media and police in public awareness of events such as riots. Together, these accounts contribute to retrospective discourses, which feed back into other accounts, such as those previously generated through mass media (see Carey 1987). The collection and examination of social media users' accounts of the events can help provide insight into how online user-generated definitions of situations now contribute to interpretations of public events, including riots.

In the sections that follow, I explore the dominant user-generated theme of criminal justice and related subthemes that emerged from posts featured on the "Vancouver Riot Pics: Post Your Photos" Facebook page. Other work specific to the 2011 Vancouver riot has examined restorative justice issues (Arvanitidis 2013), community identity construction (Lindsay 2014), collective apology narratives (Lavoie et al. 2014), and the relevance of the riot for public criminology (Schneider 2015c). The remainder of this chapter focuses on public responses to the riot on Facebook. Turning to user data generated from the 12,587 user posts made to the "Vancouver Riot Pics: Post Your Photos" Facebook group page, the remainder of this chapter looks at how online users understood the riot and its aftermath and examines the implications of these understandings for police use of social media. These posts were collected two weeks following the 2011 riot. They represent a snapshot of the overall social media response that both surprised the VPD and contributed to changes in police use of social media. These data help shape the definitions of the situation as a riot, which, in turn, influenced how information related to the documentation of the riot was constituted, perceived, and interpreted in online social interaction. Additionally, the ex post facto police solicitation of user documentation and the expanded use of Twitter helps signal the importance of this event in the growing police use in Canada of social media for investigation purposes.

The dominant criminal justice theme pertains to discourse surrounding proactive and reactive responses to the riot, with emphasis mostly upon the latter. This discourse includes user discussions of police, media, and responses by social media users to the 2011 Vancouver riot—those offered largely in response to the absence of police on Facebook.

RETRIBUTION AND REVENGE

There were clearly some users on Facebook who wished to circumvent the state-sanctioned justice system in an effort to punish those accused: "We need another riot with no police Good People vs. Bad People to the death" (June 16, 6:10 p.m.).[9] Calling for revenge, one user posted, "I am disgusted. The fucking idiots that have done this. I say start shooting them. Any pics posted here there is an automatic tag system now on Facebook . . . so they will be found out. And i want them all to get their just dessert" (June 15, 11:31 p.m.).

Most users, however, were largely in favor of strict and harsh legal punishments, such as broadening police powers to shoot suspects without due process. Posts that reiterated these kinds of sentiments were recurring: "Cops should of just shot all those pricks" (June 16, 6:41 p.m.); "This might sound bad, but if I was the swat officer I would of shot them all dead" (June 16, 10:04 p.m.). A select handful of posts also made pleas to reinstate the death penalty in Canada: "BRING BACK THE DEATH PENALTY" (June 16, 2:42 a.m.). Others called for more harsh vigilante type sanctions: "I hope all those who get charged for the riots get raped in prison and die of rectal bleeding" (June 16, 11:54 a.m.).

While such calls for "justice" were frequent, many other users called for restraint. The calls for restraint contradict the VPD's characterization of the Facebook page as a "vigilante" page. In fact, these calls for restraint appealed to the criminal justice system to punish offenders and denounced online vigilantes as "no better than the rioters" (June 20, 10:44 a.m.). One user wrote, "There is photo and video evidence being gathered for the police to decifer. The justice system flawed as it is, is still better than we as individuals gathering in our outrage and also gaining momentum on the mob mentality in the vigelantism and threats to the people involved in the rioting, and their families (inderectly). Grow up . . . " (June 20, 7:09 p.m.).

Still, as evidenced by the majority of user posts on the Facebook group page, many users indicated their belief that their online activity—posting pictures and videos and identifying alleged rioters—was conducive to official government prosecution efforts and, more generally, to police work (see also Schneider and Trottier 2012). They seemed to believe that their efforts were to assist with the prosecution of rioters, rather than in violation of existing laws, such as those that prohibit libel, threats, or harassment. One user posted, "YA PLEASE POST ALL YOUR PICS SO ITS EVEN EASIER FOR THE JUSTICE SYSTEM TO SEND ALL THERE DUMB ASSES TO JAIL!!!!!!!!!!!!!!!!!!" (June 16, 11:55 a.m.). Another user summarized the activities of the group as follows: "All this outing is strictly to allow justice under our legal system to proceed" (June 21, 12:50 a.m.); another noted, "I joined this site to help police" (June 12, 9:21 a.m.).

Many users expressed dissatisfaction with the criminal justice system, particularly delays in prosecuting those accused rioters but also its perceived leniency. One user expressed this frustration: "Maybe if there was real punishment in our justice system things would improve" (June 16, 4:20 p.m.). Another commented, "I think its great how this page and others like it, names and shames the offender. Almost better justice than the actual justice system" (June 16, 7:38 p.m.). Others were even more candid: "Really think anything is going to happen to these douche bags? Our justice system lets murderers get away" (June 17, 4:24 p.m.). According to another user, "the cops and laws are way way tolenient in Canada, these rioters need a complete ass whooping from the cops, followed by years of jail time and community service" (June 16, 4:43 a.m.).

Comments frequently juxtaposed the justice system to social media, for example, with references to the "virtual riot" and the "online mob." Users' comments indicated that they believed that their posts would have consequences that would outlast punishment levied by the criminal justice system: "Arrest them—even though with your justice system they'll be out in 3 days—let's see how long this comment actually stays on here" (June 16, 7:45 a.m.). An awareness of the youthfulness of alleged rioters and a knowledge of the differential treatment that young offenders receive in the Canadian criminal justice system were also evident: "The problem is, all these little boys will lawyer up so fast after they run home to mommy & daddy. Then our candy-ass justice system will let them all off the hook with a slap and tickle" (June 16, 3:04 p.m.).

Another user wrote of the "Vancouver Riot Pics: Post Your Photos" Facebook page, "This site is great social media = social justice with no young offenders act keep up the good work people and lets do what the courts can't, and out these losers !!!!!" (June 20, 10:48 a.m.). This post is in direct reference to accused teenage rioters and an indirect reference to seventeen-year-old Nathan Kotylak, who was identified on Facebook and other social media less than twenty-four hours following the riot. The user incorrectly references the current Canadian Youth Criminal Justice Act (YCJA), previously known as the Young Offenders Act. Under most circumstances, the YCJA prohibits publishing the name of the young offender (any individual under eighteen) or any identifying information that would identify the individual as someone who is being dealt with under the YCJA. In theory, the act applies to all form of media, including social media. However, Facebook has presented challenges to the act, including the publication and identification of Nathan Kotylak.

Kotylak's case generated discussions concerning the criminal justice system. News media widely reported that Kotylak was a member of the men's junior national Canadian water polo team and an Olympic hopeful, and he was suspended from the team shortly after he was identified online. Many expressed frustration with Canadian legislation, applauding those efforts made on Facebook to identify suspects, including young offenders such as Kotylak:

> For the ex-water polo player and all the people that became criminals over night. No scholarship, no water polo team and no Olympics. It'll be a little life the lesson that he won't forget. If he wasn't photographed would he be turning himself in? The justice system will only provide a slap on the wrist since he is only 17 so public shaming of all the people who were doing criminal acts is perfect, it will get and keep their atten-

tion. It'll eliminate the "cool factor" of what they did and show them to be just stupid, dumb, witless followers and instigators. (June 19, 12:26 p.m.)

One user wrote that he "loves the fact that a social media website is more effective at catching and also shaming criminals than the Canadian Justice System" (June 21, 6:01 p.m.). Another post read, "I hope the police do not go easy on [Nathan Kotylak] although he is a young offender" (June 18, 5:19 p.m.), and yet another suggested that "all who participated [in the riot] should be tried as adult terrorists" (June 20, 9:16 p.m.).

Not all posts on Facebook were in favor of the negative treatment of Kotylak or other suspected rioters, young offenders or otherwise. Many users referred to others on Facebook as "vigilantes" engaging in a form of online mob justice or "social witch hunt," as one user put it. One user wrote, "Why support vigilante justice? Why not now turn our attention to things that would only make Vancouver better? The social mob against these people is even sadder then the actual riot" (June 20, 8:58 p.m.). Some users went further and defined activity on the group as socially harmful:

> Also I feel like degrading these people over facebook saying they do not have an education, disgrace, calling them names, etc. is a little immature when your busting these people for rioting and such. Turn them in and let the police deal with it, cut out the immature bullshit, be the bigger people in this situation. Messaging Brock Anton over and over with bullshit is childish. The law will deal with him, dont be foolish people. (June 18, 9:41 p.m.)

Like Kotylak, Brock Anton was a suspected rioter who became a target of social media scorn shortly after the riot. Reference to Anton was a way to frame the group as susceptible to a mob mentality, having more in common with the rioters than with police. Other users took this framing even further when suggesting that individuals who were documenting the riot in order to gather evidence against suspects were actually contributing to criminal activity, a perspective supported in the external reviews of the riot (see Furlong and Keefe 2011; VPD 2011): "im sorry . . . but if you stayed and watched/filmed/pictured the rioting . . . you are part of it. the police had to deal with the 'entire' numbers . . . and the troublemakers fed off the crowd, and enjoyed the attention from 'fans' picturing their conquests. People should have left, so the police could round up the vandals quickly" (June 16, 11:48 a.m.).

While such responses were sometimes viewed in defense of the rioters, most users nevertheless seemed quite satisfied with the overall process of identifying suspects online: "I understand the VPD is frustrated with people posting too many details on here, which is causing situations like Nathan Kotylak's family going on sudden vacation due to threats. That stuff is unacceptable. *The flip side though is I've seen probably near 100 people identified on this page"* (June 20, 7:07 p.m., emphasis mine). Even when acknowledging that excessive persecution is "unacceptable," such effects are absolved in recognition that so many other alleged rioters will be identified through the efforts of users on the Facebook group page. The possibility that such persecution directed toward Kotylak and his family may spread to the "near 100 people identified" is not discussed.

POLICE AND RESPONSE TO THE RIOT

Users believed they were aiding police efforts and expected that others were doing the same. This was a dominant theme across postings. Robert Gorcak, creator of the Facebook page, wrote, "The police didn't see what happened . . . perhaps someone in this group did!" One user instructed others, "If you recognize anyone in any of the images posted here, let your local police know. These people must be held accountable" (June 15, 11:12 p.m.). This decision to share information with police is presented very matter-of-factly. If a photo or other information were available, they should be sent to the police, or, at the very least, suspects should be identified by users on the Facebook page.

Believing that they were acting on behalf of police by soliciting information, many users reposted the VPD email address, which appears 140 times in posts on the page, alongside user calls to send pictures, videos, and names. Even the *Province*, one of two daily newspapers in Vancouver, started a Facebook page to solicit information about acts of civilian heroism during the riot. Police and other emergency service personnel were often heralded as heroes:

> I would like to thank all the emergency response teams, City Police, RCMP, BC Ambulance, BC Fire Department and all citizens for giving it your all to try to end the violent acts against our Beautiful City of Vancouver it's Businesses and Citizens who were affected. To the Vancouver Police, I don't think it would have mattered how you handled the situation you would have been critized regardless. You're damned if you do and damned if you don't. Personally I believe you did the best you could with the resources you had. Hold your heads high and Be Very Proud! (June 17, 5:08 p.m.)

Users applauded police efforts, and many noted that "the police are not to blame, like it or not, they did what they had to do" (June 17, 5:31 p.m.). Rather, as one user suggested, "it is our leaders fault for not being prepared, not the police" (June 17, 5:31 p.m.). Other users posted thank-you notes to the police, and some recounted how they had personally thanked police. One user wrote that he "stopped and thanked the police officers on Granville [Street] for their awesome job during the riot" (June 20, 2:26 a.m.). In some circumstances, users even sought to identify and name emergency personnel for special recognition. One user wrote

"NOW THIS IS A HERO!! does anyone know the name of this firefighter? . . . This man deserves to be commended (as i am sure all of his squad do)" (June 19, 12:51 a.m.). Others, including mostly civilians, that "stood up" to the rioters were also described as heroes. As one user put it, "The people who stood in front of stores and tried to protect property or others should be recognized as heroes" (June 20, 7:59 p.m.).

Early posts on the Facebook page were being made while the riot was in progress. One early user remarked, "WOOHOO!! lets get these idiots arrested!!!" (June 15, 11:31 p.m.). Users claimed that police were "asking people to join this page and tag people in pics they see breaking the law if you recognise them" (June 15, 11:09 p.m.). There is no evidence to support user claims that police wanted people to participate in the Facebook group. A CBC Television news report broadcast during the eleven o'clock news on the night of the riot, however, did link the police and the mayor of Vancouver with participation on the Facebook page:

> I just quickly want to mention, to touch on something that the mayor [Vancouver Mayor Gregor Robertson] said. Anyone that is taking photos of people breaking the law, well, police want to see them. I've been told about a Facebook page that has just been started called "Vancouver Riot Pics: Post Your Photos." It's not to glorify what is happening. It is for people to post their photos, and other people to go on and, if you recognize people that are breaking the law, you can tag them with the [identifying] information. (CBC TV [Vancouver] 2011)

A few days following the riot, Gorcak noted the following in a post made on the "Vancouver Riot Pics: Post Your Photos" Facebook group page:

> Just so all are informed, your pics and help have led to arrests and is continuing to help find more people still at large. I spoke with Constable Lindsey Houghton, Media Relations Officer Community & Public Affairs Section Vancouver Police Department. He has informed me that the investigators are overwhelmingly pleased with what we are doing here, and will be in contact more later today. Thank you!! (June 18, 12:32 p.m.)

Such posts facilitate the Facebook group's apparent perception of itself as closely aligned with police and state control efforts. Prior to the end of the riot, and because of the perceived absence of police, some users complained that the police were neglectful. Messages such as, "Unbelievable. Where the heck is the police!?!?!?!?!?!" (June 15, 11:15 p.m.) characterized some users' concerns. The perceived lack among users of an immediate police response helped set the early tone of the Facebook page:

> Im So Upset That People Are Destroying Our Beautiful City And The Police Dont Seem To Be Doing Anything . . . Why, My God Why . . . What Is Wrong With You People??? Come On Police Do You Job Already . . . This Is So Crappy, Horrible, And Disappointing . . . Why Would You Want To Destroy Your Own City??? Stupid If You Ask

Me . . . Time To Grow Up . . . We Are Better Than This!!! Where's Our
Canadian Pride??? (June 15, 11:17 p.m.)

Comparing the flurry of social media activity against the perceived lack
of arrests following the riot prompted some users to encourage each other to punish suspected rioters to compensate for the perception of police
inactivity: "The likelihood of the police being able to take action of many
of these troublemakers is relatively low. However, I encourage all employers, bosses and managers to reprimand and/or fire their employees
for participating in these riots" (June 15, 11:29 p.m.).

Given that police are a key element of a riot, the fact that they were
framed as having a relatively weak presence during the riot facilitated
social media users to insert their actions in a criminal justice framework.
However, other users believed that they should let police do police work
and not interfere with finding suspects because they are not trained to do
so. Some users believed that the most they could do was send potential
evidence directly to police, as opposed to posting content on the Facebook group. Such recommendations to "let the police deal with it" were
linked to concerns about vigilantism, as explored in the previous section.

Overall, group sentiment vis-à-vis the police varied. Some members
saw the group as exceeding police capabilities; others thought the group
was interfering with police work. Yet a consistent theme was that policing, in the abstract, was a necessary good when it came to responding to
riots. Many users publicly thanked and showed appreciation for the police. This gratitude not only framed police and police work as on the right
side of judgment but also aligned the group with police work. As one
commenter remarked, "News paper article stated that 200-250 police
were on hand at the riot. Do the math 500 rioters for every ONE cop.
Under these circumstances I'd say our Police force handled them selves
to the best of their given circumstances. Kudos to the boys in blue !!"
(June 23, 5:44 p.m.). Indeed, most criticism, when it was offered, was
usually relatively mild, and often included praise. One user posted,

> I want to thank the Police Officers who did their best that night. I know
> either way you handle things you were going to be scrutinized by the
> media. If you forced the rioters with rubber bullets and tear gas you'd
> be raked for EXCESSIVE FORCE, or like they are doing now, saying
> you didn't do enough and should have pushed the rioters out of the
> city. THANK YOU for a job well done. Most of us are thankful. (June
> 17, 4:54 p.m.)

This gratitude suggests that police are designated as a necessary aspect of
a riot, even if their performance is suboptimal. Indeed, the "damned if
you do, damned if you don't" attitude suggests that both police and
criticism of police conduct are included in accounts of riots. The theme of
criminal justice allows us to understand how the definition of the situation informs collective behavior in the context of the "mediated crowd"

on Facebook (Baker 2012a). In the case of the 2011 Vancouver riot, individuals interacting on Facebook formulate the normative expectations for good behavior in relation to vigilante justice as a perceived form of police work (see Schneider and Trottier 2012).

SUMMARY

Examining responses to events on social media sites provides empirical possibilities to further develop scholarly understandings of collective behavior in relation to the definition of the situation. The formation of temporary gatherings, such as riots, has traversed into the realm of social media (Baker 2011, 2012a, 2012b), and some important changes have followed. First, the temporality of everyday face-to-face gathering has expanded online and is therefore documented and retrievable. This means that it is longer lasting; it can always be recalled and remembered by anyone with online access. Collective behavior can be (re)observed in these spaces. As outlined above, the interplay between the individual and the collective contributes to the process of meaning making and how users make sense of social situations.

Riots, as collective events, are temporary. Yet collective responses online to these events are enduring. Because of changes in time and space, Facebook enables a continuity and uniformity among users—a process that facilitates online gatherings to prolong. While online users do not collectively represent a single homogenous group, the assemblage and dispersion process remains less distinct than in the everyday face-to-face realm. Let us return to the two questions posed earlier in this chapter: What can examining the online response to social events like crisis events tell us about the meaning-making process in relation to riots? And what can activity on Facebook tell us about the social process of the definition of the situation?

In response to the first of these two questions, Facebook, as a social media platform, aids in direct user participation of the constitution of the Vancouver 2011 riot. Facebook operates as an arena for user participation, a space that does not preserve or protect anonymity. Indeed, it does just the opposite. While those in face-to-face crowds are also usually not anonymous, as evidenced in the literature on the subject, the data herein provide empirical accounts of the expectations that users have of others, including the expectation to identify, capture, and prosecute alleged rioters even beyond police efforts to do so. Social media allows the possibility for directly identifying suspects, yet, in practice, many users rely on categories and stereotypes when making such events meaningful. As the mandate to identify suspects becomes more explicit in future riots, subsequent research might examine how real-time, as well as retrospective, identification of suspected rioters impacts attempts to provide accounts of such events.

In response to our second question, the statements made by users on Facebook provide social cues for others that contribute to user understandings of the definition of the situation. Facebook is a space where users can quickly ascertain the possibilities for action. The definition of the situation sets the parameters for the possibility of action, including what and how to respond to the social event. Moreover, postings on Facebook, as evidenced above, provide empirical accounts offered by users about how social cues help to define a situation. Through the accumulation of these responses a consensus emerges pertaining to the situation that provides insight into the collective formation of the definition of the situation.

The extent to which social media impact collective understandings of riots in Canada is scarcely understood, given the relative novelty of sites like Facebook, coupled with their very recent emergence as locations of collective naming and shaming. The research presented here contributes to this developing field (Baker 2011, 2012a, 2012b) and provides greater insight into the 2011 Vancouver riot by considering the themes that Facebook users employed to better make sense of the riot, rioters, and the broader social context of the riot itself. The accumulation and analysis of "definitions of situations" provides further insight into the process of collective interpretations, the consequences of which can have lasting implications. This activity had both an immediate and enduring effect on the definition of the situation, including how interpretations of the riot continue to change more than one year following the riot. For instance, the VPD revealed that accused rioter Brock Anton, who openly bragged about rioting on Facebook, did not commit the acts he described online (Dhillon 2012b). Furthermore, this case study provides some empirical evidence in the form of user accounts that illustrate the response on social media after the riot and helps demonstrate why this response led to changes in the way that the VPD uses social media (Longley 2013).

It remains certain that "Canada's next hockey riot will be policed differently than previous ones" (Trottier 2012a, 421). The next chapter explores police use of social media as a presentational strategy. The chapter expands upon our understandings of form and content as discussed in chapter 1. The point is to illustrate how the logic of Twitter helps constitute the way that information is processed (form) and how this information as social interaction in the form of user posts or tweets (content) contributes to institutional changes in police and policing.

NOTES

1. A version of this chapter appears as an article in *Studies in Symbolic Interaction*. See Christopher J. Schneider and Daniel Trottier, 2013, "Social Media and the 2011 Vancouver Riot," *Studies in Symbolic Interaction* 40:336–62.

2. The naming of a riot remains a political designation. To avoid confusion the term *riot* is used herein to refer to the civil disorder in Vancouver in 1994 and 2011.

3. Media that initially cooperated with law enforcement requests for raw footage were "commended" in the British Columbia Police Commission report as "these tapes provided an important and useful information base" for the identification of suspected rioters. The report continued, "We should mention that there was considerable discussion and some litigation about whether media should provide their footage of the riot to police to assist them in identifying those who participated in criminal acts on the night of the riot. The tapes were delivered to police, but we have been informed that appeals may be outstanding. Therefore, we will not comment further on the matter" (1994, 44).

4. See also Tania Arvanitidis's MA thesis, "From Revenge to Restoration: Evaluating General Deterrence as a Primary Sentencing Purpose for Rioters in Vancouver, British Columbia," for a discussion of the public effort to vilify and shame offenders on social media following the 2011 Vancouver riot (2013, 7–16).

5. The reaction on social media to the 2013 Boston Marathon bombings as a "crisis-producing event" (Manning 1992) provides a recent example of community support that develops online (see Potts and Harrison 2013).

6. For an interesting discussion of George Herbert Mead's concepts on the cyber "I," "me," and digital "generalized other," see Robinson (2007).

7. For a broad review of selected developments of the study of collective behavior, see Marx and Wood (1975).

8. According to Miller, "Such technologically enhanced ties are tenuous in that one has to demonstrate a worth or relevance to be included in a network. Being an instrumental and not obligatory form of organisation, one cannot simply be born into a network and remain in it in perpetuity, one has to prove one's relevance or usefulness to be included" (2011, 201).

9. All quotations from Facebook user posts have been transcribed exactly as they appeared online. They have not been edited for proper grammar or spelling.

FOUR

Police Presentational Strategies
on Twitter[1]

With its hundreds of millions of users, Twitter is one of the most popular online social networking sites generally and among law enforcement agencies more specifically. Although police agencies around the world are active on social media, little is known about how international police departments use them (Lieberman, Koetzle, and Sakiyama 2013). The Toronto Police Service (TPS), one of the largest police agencies in North America and the largest municipal police service in Canada, has an active presence on Twitter, one of the most active among large Canadian police services. This chapter explores the TPS's use of social media following their "social media strategy" launch in July 2011, shortly after the riots in Vancouver. Between August 2011 and May 10, 2013, I collected 105,801 official police tweets and combined these data into a single 7,498-page PDF document for examination and analysis in this chapter.[2]

Despite the growing popularity of Twitter among Canadian law enforcement agencies, little research has examined police use of Twitter in Canada. In their study, which included the TPS as well as other police departments, Albert Meijer and Marcel Thaens (2013) found that the TPS's social media strategy is based on individual officer responsibility and that this particular strategy differs from other North American police departments. Therefore, we can ask, what does police activity on Twitter in Canada, and in Toronto in particular, add to our understanding of police communication more generally? To address this question, this chapter examines police presentational strategies on Twitter and reflects on how *police calls to the public* on Twitter are managed and orchestrated for organizational purposes. Thematic tweets associated with *police professionalism* and *community policing* help clarify this process.

The very first tweet was made on March 21, 2006. As of 2015, this number had amassed to an astounding 500 million tweets per day (Twitter 2015). Twitter is a micro-blogging service that allows users to post messages, or tweets, that consist of 140 text characters. Tweets can include text and reposted tweets, or "retweets," images, links to other webpages, and videos. In general, tweets are significant because they are publications of user expressions. More specifically, the significance of tweets to policing is in the impact that individual officer comments can have on public perceptions of police, and sometimes tweets affect public perceptions adversely. For instance, one "troubling," in the words of the police, online comment made by an RCMP constable in Nanaimo, British Columbia, initiated a professional standards review (*Daily News* 2010). One post made by the constable read, "How come every chick I arrest lately refuses to put clothes on and they're the ones you never want to see naked" (*Daily News* 2010). These kinds of online posts have transformed police behavior, affecting specifically what an individual officer might choose to post on his or her personal social networking account. In response to criticism to the offending posts, the Nanaimo *Daily News* reported, "The officer says he knows his posts were in poor judgment. He said he was joking among friends and did not mean to offend anyone. 'I've got a stressful job and the way I deal with it is I use humour,' said the constable. 'It's obviously pretty stupid to post that stuff on there. I didn't intend it to go out in the public'" (*Daily News* 2010).

Concerns have also grown over how the police institution itself is represented by individual officers on social media.[3] Recent department policies aimed at protecting the image of police underscore this distinction. As one of the Vancouver Police Department's (VPD) 2012 social media policy amendments warns, "The personal use of social media may have a bearing on VPD employees in their official capacity and upon the image of the VPD" (VPD, Planning, Research, and Audit Section 2012).

Users on Twitter can follow others and be followed. Following others allows users to receive and share content that collectively helps spread important information as determined by the individual user. Users can also interact with one another on Twitter where a "degree of conversationality" exists (Honeycutt and Herring 2009). Direct public conversations between users involve use of the @ symbol "as a marker of addressivity (i.e., to direct a tweet to a specific user)" (Honeycutt and Herring 2009, 1). Tweets might also include the use of "hashtags (#s) to mark tweets topically so that others can follow conversations centering on a particular topic" (boyd, Golder, and Lotan 2010, 1). Hashtags allow users to contextualize messages of interest such as news items.

Research has demonstrated that "Twitter users tend to talk about topics from headline news and respond to fresh news" (Kwak et al. 2010). According to Twitter, "40 percent of our users worldwide simply use Twitter as a curated newsfeed" (2013). Twitter is popular as a newswire that allows users to collect near instantaneous information about various world affairs; in fact, sometimes news breaks on Twitter before news media report the story. During the 2010 earthquake in Haiti, for example, some of the very first accounts and images of the destruction appeared on Twitter. While "Twitter covers most of the events that are reported by newswire providers . . . there is no evidence that one source leads the other in terms of breaking news" (Petrovic et al. 2013). Nevertheless, Twitter is now a basic news source for those seeking up-to-date information. A 2012 *Globe and Mail* report about traffic problems in Toronto helps illustrate the point: "Jim Curran is retiring this month after 40 years of reporting on Toronto's traffic for the CBC. He says he's seen Toronto's commuting patterns get so complicated that radio reports just aren't enough any more. *He recommends drivers also arm themselves with Twitter traffic updates from news outlets and municipal fire, police transit and transportation departments*" (Merringer 2012, M6, emphasis mine).

For three basic reasons, it seems like a logical step for emergency service personnel, but especially police, to incorporate technologies like Twitter into police strategies of communicating with the public. First, the use of social media complements the existing desire of police services to expand and strengthen ties with communities, including initiatives such as community policing. These initiatives involve "creating new cultures within police departments" (Skogan and Hartnett 1998, 5) that facilitate a digital culture of transparency (see Bertot, Jaeger, and Grimes 2010), which includes the "organizational decentralization and a reorientation of patrol in order to facilitate two-way communication between police and the public" (Skogan and Hartnett 1998, 5). While Twitter has the potential to meet community policing initiatives "to enable two-way discussion on policing issues" (Crump 2011, 23), it has been shown that "overall, city police departments [in the United States] do not use Twitter to converse directly with members of the public" (Heverin and Zach 2010, 6). Elsewhere, research in the United Kingdom has shown that during the August 2011 riots "police face[d] difficult problems in making effective use of social media services such as Twitter" (Procter et al. 2013a, 433).

Second, evidence suggests that members of the public now, to some degree, expect an online police presence (Accenture 2014). Some police services in Canada, including the TPS, allow minor criminal offenses to be reported online. For example, following an alleged assault in 2009 in Toronto, celebrity blogger and active Twitter user Perez Hilton posted the following tweet: "I'm in shock. I need the police ASAP. Please come to the SoHo Metropolitan Hotel now. Please." A *Globe and Mail* report on the incident continued, "A spokesperson for the downtown Toronto hotel confirmed that police arrived to take a statement from Hilton shortly after he posted on Twitter. . . . Hilton also defended his use of Twitter, saying that the Toronto police were slow to respond and he feared for his safety. 'I felt helpless and that was my very public cry for help,' he said" (Wintersgill 2009).

Third, Twitter creators modeled the service after existent police technologies (Hartley 2008). As Burlington, Massachusetts, Police Lieutenant Glen Mills, who manages the department's Twitter site, remarked in a *Boston Globe* article, "When you initially hear about it, you think, 'What a stupid idea; who's going write a 140-character message?' But it's turned out to be perfect for police. This is how we talk to each other on the radio. It's a great way to get information out quickly" (qtd. in Laidler 2014). The logic of the Twitter format follows the parameter of dispatch systems that are already familiar to police. Regarding the creation of Twitter, co-founder Biz Stone explained, "The idea came from my colleague [Twitter co-founder and CEO] Jack Dorsey, who had long been fascinated with the idea of dispatch. He used to write software for taxicabs and ambulances. . . . He wondered if the simple concept of 'status' that is so prevalent in dispatch could be applied in a social way" (qtd. in Hartley 2008, 36). This brief history at least in part explains why Twitter quickly became one of the most used social media sites among Canadian police agencies. City police forces in Toronto, Montreal, and Vancouver have active Twitter accounts. As do the Ontario Provincial Police (OPP) and Sûreté du Québec (SQ), the provincial police forces of the two most populated provinces. The RCMP also have Twitter accounts across Canada, with at least one account to represent each province.[4]

Research on police use of Twitter indicates that use involves the circulation of incident- and crime-related information (Heverin and Zach 2010) and responses to public inquiries (Crump 2011). Police also share information, such as traffic problems and public service announcements, with the public on Twitter and solicit information from the public about missing persons, for example. This research also demonstrates that the "public primarily uses Twitter to redistribute information shared by police departments" (Heverin and Zach 2010, 1). Twitter has become a contemporary way for police to manage new social "media sources of information as well as to manage information in direct citizen encounters" (Manning 1997, 7; see also Ericson, Baranek, and Chan 1987; Ericson and Haggerty 1997).

Very little research has examined police use of Twitter in Canada (Meijer and Thaens 2013), and no research has examined police officer tweets, even as police use of Twitter in Canada has grown quite dramatically and quickly since 2008 when the very first official police tweet was made from Toronto. By 2013, the number of tweets made by the TPS ballooned to more than one hundred thousand. In the United Kingdom, the growth of police use of Twitter since 2008 has been attributed to organizational changes in support of the police mandate (Manning 1978) that seeks to maintain social stability (Crump 2011). Evidence suggests that this is similarly the case in Canada, as outlined in this chapter. These changes prompt a few questions specific to police agencies in Canada: In what ways are police using Twitter? And how does the use of Twitter contribute to the development and expansion of police presentational strategies? Aside from posting crime-related information and public service announcements, what else are police in Canada doing on Twitter? Furthermore, what does this information about police use of Twitter add to our understanding of police communication?

POLICE AND TWITTER IN CANADA

Canadian police officers began appearing on Twitter in an official capacity in early 2009. Among the very first officers to use Twitter was Sergeant Tim Burrows of the TPS. As Sgt. Burrows recalls, "In late 2008, a friend of mine suggested that I should start using Twitter to connect my messages with the public. I laughed. The thing was, I had no idea what I laughed at. I had heard of Twitter but I truly didn't understand it or the power that was available in a mere 140 characters" (2014). According to Christa Miller, co-founder of weekly Twitter discussion #copchat (discussed below), Sgt. Burrows was first active in his capacity as an officer on social media sometime in late 2008. Miller writes,

> Back when I joined Twitter in late 2008, Tim was just one of the very few sworn police officers tweeting and blogging with a pioneering eye toward building a community, a virtual extension of the one he actually served. Eventually, his activity—rare among police active in social media, though thankfully less rare now—became the seed (and later, the foundation) for the way Toronto Police Service implemented social media throughout its service. (2012)

In March 2009, Sgt. Burrows launched a Twitter page "to keep Torontonians informed about the city's traffic." He was "the only Toronto police officer [at the time] using Twitter in an official capacity" (Drake 2009). Sgt. Burrows remarked, "My immediate bosses were open-minded enough to let me give it a try, so I started the Traffic Services (http://twitter.com/trafficservices) Twitter account" (qtd. in Burrows 2014). As of 2015, the TPS has the most active Twitter accounts listed under a single police force in Canada.

Sgt. Burrows also helped develop an Ontario-wide social media training program for police, having served as a part of the TPS Corporate Communications "team that developed a social media program for Service members." "Nearly 200 command and frontline officers" have been trained at the Ontario Police College (TPS 2011a).

In April 2010, Constable Scott Mills was hired as the first official TPS social media officer. He was "one of the first cops in Canada with that job description" (Allen 2011). Constable Mills was hired in this capacity in part because of his online connections with youth. According to a *Toronto*

Star report, his philosophy consists of forging "positive relationships with kids online to try to prevent crimes" (Poisson 2012).

Police services across Canada soon started hiring their own social media officers. In December 2010, the same month that the Vancouver Police Department joined Twitter and relaunched their Facebook page, Constable Anne Longley was hired as the VPD's first social media officer, as discussed in the previous chapter. The VPD joined Twitter in an effort to expand their Community and Public Affairs Section (Longley 2013), even though at the time the VPD had no existing "policy to guide employees with their personal use of social media, nor [were] the investigative and operational uses of social media governed" (VPD 2012). In fact, before 2012, the VPD had only circulated "several bulletins offering guidance to employees" about social media use (VPD 2012). Presumably for this reason, preceding the official VPD Twitter launch, Constable Longley turned to other police agencies "to find out what they did and how they did it" and noted that the "Toronto Police were a great help" (Longley 2013). VPD policy on social media now provides officers with advice on protecting their careers. For instance, they advise, "When in doubt, if you wouldn't do it in uniform, don't do it online. If you are unsure, seek the guidance of a supervisor or your union representative" (VPD 2012).

The TPS is the active leader among Canadian policing agencies in both developing and incorporating social media into existing police strategies. The TPS launched its official "social media strategy" on July 27, 2011 (TPS 2011b). On its webpage, the TPS lists a total of 164 official social media accounts, which represent individuals as well as corporate strategy and divisional policing (TPS 2015). Most of these accounts are associated with an individual officer by name. Seventeen Facebook accounts and one YouTube account are also listed. Officers are on other social media platforms as well. According to Sgt. Burrows, "[Early] it was decided that [TPS] members that received [social media] training would be authorized for Twitter and Facebook. They were also being shown how to use other platforms, and they were given an introduction to their functionality, but the Service knew that a concerted effort first on big platforms that were attracting the masses was the best place to put the social efforts" (2014).

Constable Mills (@GraffitiBMXCop), Canada's first social media officer, is one of the most active officers on Twitter in Canada, having posted more than forty-four thousand tweets. Constable Mills is also on other social media sites, including Google Plus and YouTube. However, only Twitter and Facebook icons are listed next to his name on the TPS social media page. Links to officer social media accounts on the official TPS webpage suggests endorsement of officer activities by the TPS.

POLICE PROFESSIONALISM

Like all Canadian police departments, the TPS has a Professional Standards Investigative Unit whose purpose is "to contribute to the achievement of the Toronto Police Service's mandate" through "the promotion and support of professionalism" (TPS 2013a). The police mandate includes upholding the appearance of the police as an apolitical organization (Manning 1978). In single tweets and Twitter conversations with others, officers discuss many topics, ranging from professional to personal issues. Reflecting on the Occupy Toronto protests in November 2011, Superintendent Neil Corrigan of the TPS Professional Standards Investigative Unit tweeted, "Great day yesterday at Occupy Toronto, our officers conducted thems with dignity and professionalism. I am proud of them" (November 24, 2011, 5:01 a.m.).[5] Use of "professionalism" by police administrators like Superintendent Corrigan is one way to distance the institution from any untoward appearance related to the influence of politics (Klockars 1988).

Officers of all ranks making any explicit political statements on Twitter, such as those that might be interpreted as an endorsement of political candidates, could be viewed as a violation of police professionalism or, worse, may even constitute an offense. The Police Services Act of Ontario prohibits police officer endorsements for or against a political party or candidate *when on duty*. Official police use of Twitter and other social media expands the parameters of police work to also include presentation strategies such as maintaining an apolitical appearance while off duty. This development is advanced by the communicative format of Twitter that allows officers to actively make official statements 24/7 as an officer (i.e., on behalf of the policing institution), both while on and off duty, and often during times when no media statements were needed, requested, or even required.

In the examined data, police tweeted (some regularly) on official TPS social media accounts *while off duty*. A handful of these tweets had little or nothing to do with police work—that is, crime work (Ericson 1982). A tweet on an account that has since been deleted made by Staff Sergeant Daryle Gerry read, "on vacation for 2 weeks, that includes twitter just golf—sun—and idleness" (May 22, 2012). Another tweet by TPS officer Detective Jeff Banglid read, "Vacation time in fall is fantastic! Off to play some golf. Hopefully I break 70" (October 18, 2012, 7:02 a.m.).

It is not clear if the Police Services Act or other regulations are applicable to those tweets, political or not, professional or not, made by officers on authorized TPS social media accounts while off duty. Analysis of officer tweets and interaction with other officers and members of the public on Twitter in fact suggests that police officers have an explicit desire to maintain an apolitical appearance, a feature of existent police presentational strategies (Manning 1978, 1997). As one TPS tweet reads,

"Our duty is to enforce the law, not impose our politics. There must be a separation" (@The22News, July 4, 2012, 6:21 p.m.). This presentational strategy is further illustrated in "#copchat."

According to co-founder Sgt. Burrows, #copchat is "a weekly conversation that takes place every Wednesday night starting at 9pm Eastern, on Twitter" (Walking the Social Media Beat 2015). The very first #copchat occurred on June 27, 2012. The "chats" last for an hour and regularly include TPS officers. As explained on *Walking the Social Media Beat*, #copchat's website host, "Each week we discuss a different topic from the world of policing, law enforcement and other like-minded industries as it relates to the use of social media, Internet, communications, community building, operations and investigations" (2015). The topics can be preselected or may emerge in situ. Members of the public are encouraged to participate. The same officers often participate on TPS accounts each week, even while off duty, or on vacation. One officer tweeted that he was participating "from the hot tub" (October 10, 2012, 6:47 p.m.).

A November 14, 2012 #copchat addressed the issue of politics. The 22News (@The22News) tweeted, "Politics and Police, this should get interesting!!" (6:07 p.m.). The22News represents the 22 Division that provides service to the western suburban region of Toronto. Three minutes later, @The22News tweeted, "Police should not individually identify themselves with political parties" (6:10 p.m.), and followed up quickly with, "An association or police union can speak about politics" (6:14 p.m.). A discussion between The22News, Sgt. Burrows, and Constable Mark Earl (the TPS School Resource Officer) ensued over individual officer expression of politics.

It was agreed that off-duty politics is acceptable but not in communities in which officers reside. In such circumstances "a lawn sign" or "bumper sticker on your car" supporting a political candidate would not be appropriate (Sgt. Burrows, 6:13 p.m.). Later that evening during #copchat, Sgt. Burrows reiterated his view on politics by simply tweeting, "we're not allowed to talk about them" (6:49 p.m.). TPS Constable Rob McDonald elaborated, "to deny that we have them would be wrong" (6:51 p.m.). However, he later added, "we must acknowledge like other professions we have them and unlike other professions we shouldn't xpress them" (7:03 p.m.).

A previous exchange between these two officers indicates an expressed interest in, albeit not an endorsement of, partisan politics. On August 25, 2012, Sgt. Burrows tweeted to Constable McDonald that he was in Florida. Constable McDonald responded with, "Wow so you are adding stormchaser to your resume or are u that big of a #US politics junkie to go to a #Republican Convention" (6:32 p.m.). Sgt. Burrows replied, "Both, but more interested in the politics" (August 26, 2012, 7:21 a.m.). This statement does not necessarily express support for one political party, but it nevertheless appears suggestive of more than a passing interest in a U.S. political party with a well-publicized "tough on crime" agenda (Becket and Sasson 2004). Sgt. Burrows expresses this interest again months later during a #copchat discussion: "Doesn't it make sense for the police to support a politician that is tough on crime?" (6:27 p.m.).

Sgt. Burrows is the author of the thirty-nine-page *Walking the Social Media Beat: Twitter Guide for Police and Law Enforcement*. The book, according to his website, is "the 1st and only book made specifically for police and law enforcement on the use of Twitter" (Burrows 2012a). In the preface of the book, under the heading "Why a Twitter Book Specific to Police and Law Enforcement?," Sgt. Burrows writes,

> Police are held to a higher standard as they should be and that includes the way in which you use the tools that are available. Although we all have the ability to and protection from law to say what we want, we also have a responsibility to ensure we are always professional and above reproach. A citizen can freely give an earful to a politician or call someone out for their behavior online or in real life provided they don't break any laws, but as a police officer, law enforcement organization *or the public face of an agency account you don't enjoy that degree of freedom.* (2012b, 6, emphasis mine)

Political endorsements on agency accounts were not present in the examined data. Officers did retweet politically themed tweets, but they were usually nonpartisan news reports with headlines like "[Jim] Flaherty intends to stay Finance Minister until Ottawa's books are balanced." When these and similar (re)tweets were made, they were often politically inconsequential. An example includes a TPS retweet of a tweet made by the prime minister of Canada: "Wishing the Canadian Olympic Team the best of luck in London! #Olympics" (July 27, 2012, 8:17 a.m.). Retweets of this variety, such as those made by political figures, were uncontroversial and, more importantly, nonpartisan. In other possible political circumstances, officer tweets directed attention away from political issues. On one occasion, one of the highest-ranking officers of the TPS, Deputy Chief Peter Sloly, retweeted, "Lawful Access Appeal—This is not about Politics, it's about public safety!!!" (October 26, 2012, 10:32 a.m.).

Officers also tweeted about strengthening laws against crime. An April 21, 2013, retweet by TPS Constable Mark Earl made by then Canadian Prime Minister Stephen Harper illustrates the point: "Our govt recognizes the impact that violent crime has on victims. That's why we are creating a Victims Bill of Rights." The retweet included a link to a government press release about the Not Criminally Responsible Reform Act, which seeks to "increase the efficiency of the justice system [and] strengthen laws against perpetrators" (Prime Minister of Canada 2013). The government announcement of the act (Bill C-54) was met with partisan fervor. A CBC News report illustrates the point:

Justice Minister Rob Nicholson testified at the committee on Monday that the provinces support the proposals and that he's talked to many victims, *but* opposition MPs have been raising questions about who else was consulted in the drafting of the bill. On Wednesday, many of the witnesses, including representatives of mental health groups and criminal justice organizations, said their input was not asked for in the drafting of the bill. (Fitzpatrick 2013)

Possible partisan connections, such as an endorsement in the form of a (re)tweet about, for instance, tough-on-crime legislation as a way to address mental illness concerns, can more readily be managed, and any subsequent criticism can be deflected away from police through officer interaction with members of the public on Twitter. An officer–public interaction on October 10, 2012, immediately following a tweet about mental illness, helps further illustrate the point. Deputy Chief Sloly tweeted, "It's World Mental Health Day—many people in criminal justice system suffer from mental illness. More awareness = less crime/more justice!" (9:10 p.m.). A citizen responded, "many of those that suffer mental illnesses are young black men. They suffer because they are stigmatized and criminalized" (10:08 p.m.). Here the citizen reframes the issue as one associated with race. Deputy Chief Sloly "agreed" with this citizen assessment, then redirected the blame: "All communities/cops need to better understand mental illness impacts re criminal justice system + general society" (11:24 p.m.). Police–public communication occurs frequently on Twitter. Notably, police–public interactions are a basic thematic often associated with various community policing initiatives (Skogan and Hartnett 1998). In other words, the degree to which police–public relations involve partisan politics and the process by which police–public relations contribute to developing community policing initiatives are central concerns of police presentational strategies.

TWENTY-FIRST-CENTURY COMMUNITY POLICING ON SOCIAL MEDIA

The use of social media sites has "emerged as a valuable tool for proactive police communications strategies . . . allowing for more direct and intimate connections with a public whose approval they so desperately seek, [and providing] a new conduit between police and the public" (Lee and McGovern 2014, 117). An examination of the TPS social media strategy reveals the rhetorical circumlocutions of community policing (Klockars 1988), including decentralization, reorientation of patrol, and increased communication between police and citizens. According to a TPS report, "the new [social media] strategy allows the Service to enhance public safety, public service and public trust while improving internal communication and reaching out to an extensive audience" (TPS 2011b). In the same report Deputy Chief Sloly explains,

> The biggest change for us (with the new technology) is our culture. We are not used to this type of *decentralized*, high-speed, *highly interactive information-sharing environment*. Traditionally, policing is a very hierarchal and para-military culture. We don't give our frontline people a lot of opportunity to speak on behalf of our organization. This is changing all of that and because of that radical change, it made people like me very nervous. It took a lot of convincing and once the lights went on in my head, the lights went on right across the police service. What I love about this approach is that it's bottom up. It's our frontline people— those responsible for public safety and *interacting with the public*—that are telling us that this is the way we need to go. (TPS 2011b, emphasis mine)

Police use of Twitter accomplishes the implementation of two-way communication that more readily "encourages information exchange" (Skogan and Hartnett 1998) with the citizens, while also serving as a highly visible online public police presentational strategy (Manning 1978). According to Sgt. Burrows,

> engagement [on social media] was a top priority of the program. If someone had a question, you answered it. Needed information? You got it. Plus, the members and units had the autonomy to make their voice in the program unique and individualized to heighten the relations that were built. . . . In the beginning I was told by one person that using social media wasn't a very good idea because, as he put it, "What will you do if someone asks a question?" That was the ultimate risk aversion statement I had ever heard. Risk management is not only about managing risk, but it also embraces taking calculated risks that provide measureable benefits. (2014)

These initiatives include developing social media like Twitter to publicize police social activities, including activities such as officer vacations or officer sports affiliations not usually associated with the organization. This communication strategy further expands police decentralization efforts across Toronto (and the world), allowing police to become more personally entrenched in the lives of ordinary citizens and communities. In the words of Deputy Chief Sloly, the use of social media "has expanded the reach of the Toronto Police service broadly to communities that didn't trust us and wouldn't communicate with us in the ways that we were traditionally doing it" (TPS 2011c).

At first blush, police use of social media provides a legitimacy conundrum. On the one hand, the legitimacy of modern police relies on the micro-level (Ericson 1982). The use of micro-blogging services like Twitter seems to fulfill this aim. Official police tweets, however, as police "organization products" (Manning 2003b), expand parameters across time, space, and jurisdiction in which ordinary citizens might make judgments about police work (Skogan and Frydl 2004). This might include what constitutes, according to public perception, the appropriate use of Twitter by active police officers. For example, the Belleville Police Service (BPS) in Ontario launched an internal investigation after the service was alerted to "inappropriate comments" posted to Twitter by someone identifying himself as a BPS officer. The BPS also confirmed in a statement that the posts "'may' be linked to one of its officers" (Fox 2014). Tweets included, "Sometimes I wish the cruiser played 'Move Bitch, Get Out The Way,' [a popular song performed by rap music artist Ludacris] instead of using sirens" and "I like big busts and I canot lie [a parody of a song performed by rap music artist Sir Mix-a-Lot]. I'm a Cop . . . get your mind out of the gutter" (Fox 2014).

On the other hand, at the heart of police legitimacy rests "impersonal authority [which] proved to be the secure foundation of police legitimacy" (Miller 1975, 84). Social media accounts attached to individual officers could personalize police officers, which might result in the erosion of impersonal authority. Or personalizing officers, as a strategy, might more fully foster police–community relations. It becomes a worthy endeavor, then, to explore statements (i.e., tweets) offered by "frontline people" (i.e., individual officers) who are authorized to speak on behalf of the policing institution as a modern presentational strategy of a "new" police–public dynamic.

TPS officers interact with members of the public (including their friends and families) often and sometimes far beyond the scope of policing activities, during private, off-duty times and from within private spaces, such as their homes. While police may keep highly visible public discussions of their families apart from work activities (perhaps due to safety concerns), this did not seem to be the case with the TPS officers on Twitter. Responding to a "Happy Easter" tweet addressed to an officer by her first name, the officer replied, "Just finished up our family dinner. Hope you are enjoying it too" (March 31, 2013, 8:24 p.m.). Other TPS officers also tweeted about their family time from home, and other places, such as the zoo. For instance, Constable Randall Arsenault tweeted a photograph of himself holding a baby goat and wrote, "Great day spent with family at the Bowmanville Zoo. I'm very thankful for this time" (July 28, 2012, 8:38 p.m.).

These kinds of personal tweets were not uncommon. In fact, many officers seemed to encourage communication from the public during their off time. Some were occasionally apologetic for delayed responses to public queries while they were off duty. One officer tweeted, "Holiday Season 'Vacation'—Keeping in Tweeter Touch" (December 7, 2012, 8:49 p.m.). Another wrote, "Hello, sorry for the slow reply, was on holiday" (September 19, 2011, 1:16 p.m.). One officer appeared to have posted a family photo. His tweet read, "3 excited bluejays fans enroute to watch the Boston Red Sox lose to Toronto" (April 30, 2013, 1:18 p.m.) and included a photograph of three children. While no identifying information was included in the photograph, two children bore a strong physical resemblance to the image associated with the officer's Twitter profile. In fact, family and sports were frequent off-duty themes among select TPS officers' tweets. For instance, Constable Wendy Drummond tweeted the following in response to a citizen tweet: "Family time and the #leafs game = perfect weekend" (April 20, 2013, 5:08 p.m.).

Tweeting about sports, but especially hockey and the NHL Toronto Maple Leafs in particular, was another strategy officers used to solicit citizen interaction or communication. For example, Constable Drummond tweeted, "Satellite just went down . . . ugh . . . what's the score??? #GoleafsGo" (April 8, 2013, 6:32 p.m.). The tweet received eight unique responses from citizens in just nine minutes, including these: "still 4–3 Toronto" and "satelitte in and out here in muskoka too." These tweets followed immediately in the same minute as Constable Drummond's tweet at 6:32 p.m. Another response just two minutes later indicated that the user, a paramedic according to his or her Twitter profile, sent a private text message to Constable Drummond's cell phone: "Just bbm'd [Black Berry cellphone messaged] u the score . . . 3 mins to go in the game" (April 8, 2013, 6:34 p.m.). In another circumstance, Detective Jeff Banglid tweeted to a follower, "Any predictions for tomorrow's game?" (January 13, 2013, 9:18 a.m.), which elicited, "I say Leafs 4 Habs [a moniker for the NHL Montreal Canadiens] 2" (9:24 a.m.).

Sports tweets and related interaction with the public as a presentational strategy portrays the image of a down-to-earth, "average Joe" officer who shares the interests of community members. These kinds of tweets diminish perceptions of authoritarian relations traditionally associated with police. Numerous TPS officers tweeted text and photos while watching sports games both from home and elsewhere, including the stadium. When tweets from the stadium or arena were posted, officers would usually clarify that they were attending the game as a fan, not working as an officer. This talk is another rhetorical move that distances police officers from their authoritarian roles. For example, in response to a public question about his tweet, "I'll be at the Leafs game tonight!! Can't wait!! First time in the arena this year for me" (April 27, 2013, 9:47 a.m.), Detective Banglid replied, "thankfully, I'm not working. Going there as a fan" (10:04 a.m.). Even officers that were not fans used sports to interact with members of the public. In response to a public tweet, "I take it you're not that big of a hockey fan Laurie?" (January 21, 2013, 3:25 a.m.), TPS Crime Analysis Constable Laurie McCann noted, "not a huge fan of hockey :-(but I will watch it and I support the leafs!" (January 21, 2013, 3:40 a.m.). Addressing officers by their first names (which seemed to be an acceptable, if not encouraged, practice) further contributed to a visible semblance of non-hierarchal police interaction with members of the public.

Personal tweets—that is, tweets that were not police work–related—from officers were quite common. These tweets built upon previous efforts intended to legitimize police–community relations by further cultivating police officers as personable and relatable individual members of the community (see Cain 1973). Three different officers posted pictures of their dogs with, or dressed in, Toronto Maple Leafs apparel, for example. The text of one of these tweets, made by Staff Sergeant Chris Boddy, read, "My dog is pumped for the @MapleLeafs playoffs run" (April 28, 2013, 1:12 p.m.). Other personal tweets from within the parameters of the workplace were also present. Various non-monetary bets, for instance, were waged between officers and sometimes others, including a paramedic, over hockey matches on Twitter. These interactions present officer camaraderie and workplace culture as relatable social activities. These tweets, as well as other humorous tweets, also help to present officers to the public as personable and fun.

For instance, on one occasion Constable Arsenault tweeted, "That awkward moment in a Tim Hortons [coffee and donut shop] line-up, in uniform, ordering a coffee only with all those sweet, chocolatey donuts staring at me" (September 26, 2012, 6:59 a.m.). An aspiring police officer, according to his Twitter profile, responded, "Go for the Boston Cream!" (7:31 a.m.). Some officers even played up the cop-donut stereotype. Sergeant Aly Virji tweeted, "What's up with all this talk about muffins? What happened to the cops and donuts stereotype? ;)" (October 1, 2012, 7:19 p.m.). On another occasion, TPS School Resource Officer Constable Mark Earl tweeted to Constable Arsenault: "I'll even treat you to a coffee and a delicious pastry of your choice," to which Constable Arsenault replied, "what? You mean donut? What makes you think I like donuts? Let's meet up this week, I'll inbox you" (January 15, 2013, 6:55 a.m.). "Inbox you" refers to a private message between Twitter users. Tweets of this sort suggest that officers were fully aware that their messages were public.

During off-duty times, officers continued to appear interested in interacting with community members, even when presented with confrontational tweets. For example, one citizen tweeted, "we as 'troubled youth' don't trust these people in authority because it seems like all they care about is their cheque" (July 19, 2012, 4:42 p.m.). Staff Sergeant Shawna Coxon replied, "Hmmm . . . I'm at home, on my own time watching and listening bc [because] I care. I want u to know your voice is important" (4:51 p.m.). The citizen then retweeted Staff Sergeant Coxon's response, which might suggest approval or appreciation.

In another less common instance, a self-proclaimed "anarchist," according to his or her Twitter profile, added the following to a September 5, 2012 #copchat discussion: "FUCK THE POLICE" (6:13 p.m.). Debate ensued between officers and citizens, including one tweet that simply read: "Don't. Talk. To. The. Police." Constable Michael Pagniello responded, "why not? We are here to talk to the public." Sgt. Burrows added, "Why not? You have problems with the police, who do you think is going to solve them?" The anarchist followed with "because you don't protect us, you protect the rich and rape our rights." Constable Pagniello dismissed this and other evocative tweets: "How about asking some legitimate q's?" To which a citizen tweeted, "how about some legit cops . . . are there any?" Constable Pagniello then reiterated his earlier position: "There are many legit cops. Plenty here to chat with. But have to B legit Q's." Police–community relations, on Twitter, as this exchange illustrates, occur on terms framed by officers. This process allows police to effectively maintain control of the situation while also simultaneously presenting themselves as openly available to community members.

SUMMARY

Following the 2011 riot in Vancouver, police across Canada have increased their presence on social media sites, especially Twitter. Police have actively used news media to garner symbolic support of the public to encourage police efforts (Ericson 1982; Doyle 2003). A more recent and underexplored development includes police use of social media as an interactive platform to encourage symbolic support. Officer tweets and interactions with the public and other officers provide opportunities to develop and facilitate existent police presentational strategies. An important development in this area includes police use of authorized Twitter accounts in off-duty spaces, thus expanding presentational strategies to include matters not related to police work. Thematic issues associated with *police professionalism* and *community policing* were identified and explored. Let us now briefly revisit the first question posed at the outset of this chapter: In what ways are police using Twitter in Canada? To address this question, we turn our attention to our first thematic: police professionalism.

Officer tweets maintained a public professional appearance that included avoiding any discussion of explicit political views on Twitter. Officers also avoided political discussions while off duty (e.g., on vacation or at home) on authorized TPS social media accounts. TPS officers did, however, on occasion tweet or retweet politically themed content. Any semblance of partisan endorsement in such circumstances can be effectively managed through officer interaction with citizens. On the one hand, TPS officers use Twitter to make explicit claims of political neutrality; on the other hand, they use Twitter to implicitly support laws and political statements related to police work—that is, crime work (Ericson 1982). Regulations forbid police from political endorsements while on duty. It remains less clear whether this applies to police in off-duty spaces tweeting on authorized Twitter accounts, which blurs the distinction between public–private meanings associated with police professionalism in unforeseen ways. Future research might further explore this ambiguity.

Our second thematic, community policing, also provides some insight into the question of how police are using Twitter. Community policing promises "new" police–public dynamics, including two-way communication with members of the public (Crump 2011; Skogan and Hartnett 1998). The interactive format of Twitter permits police agencies to develop this strategy in visible public spaces (Crump 2011). Twitter also allows police agencies to expand decentralization efforts across service areas and include an array of subject matter (e.g., sports). These tweets were successful in soliciting responses from the public. Such tweets present officers as "an average Joe" with related community interests. These tweets help humanize officers (Meijer and Thaens 2013). In addition, tweets about officers' family time or those that show a self-referential humor, along with public tweets addressed to officers using their first names, help contribute to the process of community acceptance of police through an online version of "easing" behaviors (Cain 1973). In other circumstances, such as confrontational tweets (e.g., "fuck the police"), officers reframed the conditions of interaction to meet their own objectives. It remains to be seen if police use of Twitter and social media in a personal manner, such as those in which officers proclaim allegiance to a hockey team, will erode police legitimacy efforts that rely on impersonal authority. Research in the future might explore developments in this area.

Let us now briefly revisit our second research question: How does the use of Twitter contribute to the development and expansion of police presentational strategies in Canada? A basic finding that runs concurrent in each identified thematic concerns the official officer use of Twitter while off duty. Some of the examined off-duty tweets had little or nothing to do with official police work. Yet the appearance of these tweets made on authorized accounts nevertheless seems to at least endorse activities such as vacations, hockey games, and family time as a form of police work. The empirical evidence also reveals the expansion of police actives to include circulating information about the private lives of officers in the public realm. This process delineates the individual officer from the police organization but does so in an official capacity as a police officer, an initiative that diminishes the appearance of authoritarian relations.

Now to return to our final question: What does this add to our understanding of police communication on social media? Twitter and social media more generally represent the most recent communication technology that turns *all* police matters, including even those private and off-duty matters, into "organizational products" (Manning 2003b). Previous work has examined the significance of the "quality of routine police–citizen encounters" (Skogan 2005, 299) and how technologies affect calls to police in the way that calls are interpreted, managed, and handled (Manning 2003b). This chapter advances this police communication work in redirecting our attention to how *police calls to the public* are managed and orchestrated for organizational purposes. On the one hand, police use of Twitter represents an informal attempt at "bureaucratic propaganda" using social media that, unlike "traditional propaganda," targets "an individual, group, or specific segment of the population" (Altheide and Johnson 1980, 14). On the other hand, this attention also highlights the promotion of trust in a "risk society" (Ericson and Haggerty 1997). Social media platforms such as Twitter enable and encourage citizens and police alike to publicize their private lifestyles in ways that increasingly serve to advance surveillance relations. In their book *Policing the Risk Society*, Richard Ericson and Kevin Haggerty write,

> Risk communication systems are entwined with privacy and trust. The more foreboding and fear lead people to withdraw from public involvement, the more they value privacy and withdraw into privatized lifestyles. The greater the privacy, the greater the need for surveillance mechanisms to produce the knowledge necessary to trust people in otherwise anonymous institutional transactions. Paradoxically, these mechanisms intrude on privacy and are a constant reminder of the uncertainties of trust. Yet it is only in a framework of trust that patterns of risk can be adequately institutionalized and form the basis of decisions. Privacy, trust, surveillance, and risk management go hand in hand in policing the probabilities and possibilities of action. (1997, 6)

This chapter contributes to the developing research on police use of Twitter, provides some insight into what Canadian police are doing on Twitter, presents empirical evidence that police actively use authorized Twitter accounts *while off duty*, and contributes to advancements in our understandings of police image work on social media in Canada (Schneider 2015b). Future work may address risk management as it pertains to individual officer use of Twitter, such as when police tweet about family time seemingly without regard for personal privacy, or even perhaps safety. Exploring the gendered dimension of officer relations on and use of Twitter might provide another avenue for analysis. For example, do female officers tweet more about family and less about sports? Do police tweets reinforce policing as a distinctly masculine domain? To what extent might officer tweets reinforce conventional gender norms?

The next chapter details how the logic of YouTube constitutes the way that information is processed (form) and how this information in social interaction in the form of user posts, videos, and images (content) contributes to "crime stories" or stories that involve police (Chermak 1995).

The chapter directs our attention to reflect upon how "crime stories" can now unfold on social media ahead of police interpretation and control.

NOTES

1. A version of this chapter appears as an article in *Policing and Society*. See Christopher J. Schneider, "Police Presentational Strategies on Twitter in Canada," Policing and Society: An International Journal of Research and Policy 26, 2 (2016): 129–47, doi:10.1080/10439463.2014.922085.

2. Tweets were collected from all active TPS Twitter accounts. At the time of collection, Twitter only allowed access to the most recent 3,200 tweets. In circumstances of excessive activity, only the most recent 3,200 tweets were retrieved for analysis.

3. These concerns also extend to fake Twitter accounts that purport to officially represent police (see CBC 2011a; McGovern 2011).

4. Canada's national police force, the Royal Canadian Mounted Police (RCMP), must have social media (e.g., Twitter) accounts active in both English and French, the two official languages of Canada.

5. All tweets have been transcribed exactly as they appeared on Twitter. They have not been edited for proper grammar or spelling.

FIVE

Police Caught on Camera

Framing the Death of Sammy Yatim

The final case study focuses on a bystander video, uploaded to YouTube by Toronto architect Martin Baron, of the shooting death of eighteen-year-old Sammy Yatim by Toronto Police Constable James Forcillo just after midnight on July 27, 2013 on an empty Toronto streetcar. The chapter explores how social media framed police use of force before police themselves were able to control the definition of the situation, despite their use of social media in relation to the incident. In the words of Chief William Blair of the Toronto Police Service, "Social media can change the way we engage—it can help get a conversation going with the police and make citizens part of it. Social media and apps can allow the police to speak directly to the public and present the facts without others spinning it. We need to move with the times and we are running to catch up" (qtd. in Accenture 2013, 7).

This chapter contributes to scholarship in Canada on police "image work" (Ericson 1982; Mawby 2002b) and, in doing so, helps develop a relatively under-researched area of social media's "ability to stir the attention of both the traditional media and policy makers" (Milivojevic and McGovern 2014, 34; see also Campeau 2015; Nolan 2014; Schneider 2015b). Recent research with police agencies in Canada reveals that social media platforms have "substantially impacted the consciousness and the behaviour of operational police officers, [including] how they use physical force in their interactions with citizens" (Brown 2013, 1–2; see also Campeau 2015 and Brown 2015). This research suggests a decline in public confidence and trust in the police in Canada *because of* scrutiny by way of a "new visibility" on social media (Thompson 2005; see also Council of Canadian Academies 2014). Largely missing from this body of quickly growing scholarship is an empirical examination of this process as it unfolds and develops on social media. Gregory Brown's (2013) conclusions are drawn from surveys of 231 frontline police officers in Toronto and Ottawa, follow-up interviews with 20 of these officers, as well as interviews with 8 policing officials in the aforementioned cities. While this work is significant and provides insight into police work, neither Brown (2013) nor his police respondents conduct any analysis of social media posts. With some exceptions (e.g., Nolan 2014; Schneider 2015b), much of the scholarship in the area of police and social media consists mostly of officers merely reflecting on their own perceptions of the implications of social media on policing and police work. This chapter provides an empirical examination of social media posts offered in response to a video recording uploaded to YouTube of police use of force that resulted in the death of a Toronto teenager.

MEDIA AND POLICE USE OF FORCE

Police use of force is a basic feature of police work (Bittner 1970; Westley 1970). The question of force as justified or not is based entirely on the social context in which force is used. This can include the legal justification of the use of force. "[Egon] Bittner nor the other researchers who have focused on the use of violence by police (Westley 1970; Muir 1977; Klockars 1985) speak from a legal standpoint" (Brodeur 2010, 293). However, "the written rule is clear: cops are to use no more force than is necessary to subdue a suspect" (Skolnick and Fyfe 1993, 13). The contemporary meaning-making process (i.e., social context) regarding public understandings of police use of force increasingly occurs on social media (Nolan 2014). How materials are framed in these spaces has been shown to influence the interpretation of police use of force by the public (Nolan 2014) as well as the decision by police officers to use force (Brown 2013; Campeau 2015). This chapter examines the public interpretation of police use of force as it unfolded and developed online in the days following the on-duty police shooting of Toronto teenager Sammy Yatim in July 2013.

The justification of the use of force in the not so distant past usually relied on accounts provided by responding officers based on direct observation or experience (Goldsmith 2010). As Marvin Scott and Stanford Lyman explain, "An account is a linguistic device employed whenever an action is subjected to valuative inquiry. Such devices are a crucial element in the social order since they prevent conflicts from arising by verbally bridging the gap between action and expectation" (1968, 46). Mass media have irrevocably changed the process of police visibility and accounts, especially those accounts that relate to the use of force (Lawrence 2000; Skolnick and Fyfe 1993). The introduction of television—a medium capable of broadcasting images of policing and police work—greatly expanded the "secondary visibility" of police—that is, how police behavior appears in its representation rather than firsthand interaction (Goldsmith 2010). The dissemination of images of police force to large segments of the population can present the use of force (justified or not) beyond the immediate control and interpretation of police, which may result in unintended and negative consequences for police. The 1968 National Democratic Convention in Chicago, Illinois, serves as a prime example in which televised images revealed the commonplace use of violence as a basic feature of "the culture of policing" (Stark 1972). During the convention, television footage showed the use of "indiscriminate violence" as police beat protestors, activity characterized later in a commissioned report as a "police riot" (Walker 1968).[1]

In the 1990s, police "image work" increased in significance (Mawby 2002b). This shift can be attributed to the expansion of recording technologies that increased the visibility of police use of force to the public (Goldsmith 2010). The 1991 citizen-recorded video of the police beating of motorist Rodney King in Los Angeles served as the watershed moment of media exposure of police violence (Skolnick and Fyfe 1993). The recording was "seen everywhere in the world" and was, at the time, "the most explicit and shocking news footage of police brutality ever to be seen on television" (Skolnick and Fyfe 1993, 3). Perhaps even more remarkable, George Holliday, the man who filmed the beating, had originally intended to provide his recording of the incident to police but, because an officer reached by telephone seemed "so uninterested," Holliday arranged to provide the tape to news media instead (Skolnick and Fyfe 1993). The video footage and subsequent acquittal of the police officers involved in the beating of Rodney King would spark some of the worst mayhem in U.S. history.

The beating, however, "was unusual only because it was videotaped" (Williams 2007, 6). Police violence has not necessarily increased, but recordings of this violence certainly have. A cursory search of the term "police brutality" on YouTube returns nearly twenty-eight million video clips (Brown 2013). Returned results include footage from mobile phones, surveillance cameras, and even police dash cams. Other videos include excerpts of television news reports. Video views span from just a few to millions. One video titled "POLICE BRUTALITY—Recording the Police is Dangerous, but Necessary!" is a compilation of user-recorded footage with a voiceover narrative that encourages citizens to film police–citizen interaction. The video has been viewed more than two million times.

Police officials actively work to construct positive images for presentation in news media (Ericson, Baranek, and Chan 1987; Fishman 1981), including perceived police effectiveness (Dowler 2003) to control "the space and create the story before anyone else does" (Lee and McGovern 2013, 119). "The news-media is pivotal to the ability of authorities to making convincing claims" (Ericson, Baranek, and Chan 1991, 8), much of which includes claims about crime and crime control (Ericson 1982; Manning 1978). Studies suggest that accounts of crime are an important and central feature of news media (Graber 1980; Surette 2010) and that police "often exert substantial control" over crime accounts (Doyle 2006, 870). A significant percentage of crime news stories appear in the "beginning stages" of the criminal justice process where "police decide how these recently discovered crimes get presented because minimum information is available from other sources at this stage" (Chermak 1995, 33). Research suggests "up to two-thirds of people find out about the police through the media rather than personal contact (Skogan 1990:18–19; 1994:13–14)" (Mawby 1999, 271). Social media platforms such as YouTube now contribute to this process.

CRIME STORIES ON SOCIAL MEDIA

Research in Canada has examined police accounts presented in news media in response to discrediting materials on YouTube (Nolan 2014; Schneider 2015b). Compared to the era before social media, police now have much less control over how some crime materials are obtained, presented, and framed by online users and in news media, as chapter 3 illustrated (see also Nolan 2014). Journalists are also no longer forced to rely strictly upon police officials as primary information sources. This is not to suggest that police are less relevant as providers of official information, as "authorized knowers" (Fishman 1980). Research has illustrated that police can gain control as stories progress in the news cycle by providing official information produced by police themselves (Doyle 1998, 2003; Schneider 2015b).

Chapters 2 and 3 addressed police and citizen responses to crime as it unfolded concurrently on social media. Research has yet to examine the impact of the production of crime stories, such as those that involve police shootings, that *begin* on social media. This chapter seeks to understand the development of this process. The concern here is what the phenomenon whereby stories begin on social media can tell us about the relations between communication formats and institutionalized control — that is, institutional legitimacy or the judgments that people make about the rightfulness of police conduct (Skogan and Frydl 2004). These judgments can now draw on materials circulating on social media. Looking specifically at the Yatim case, I argue that materials published on social media provided the dominant frame of the death of Toronto teenager Sammy Yatim, including how his death was quickly understood and interpreted *largely in absence of police accounts*. This is unusual for two reasons. First, the event involved a crime story; Yatim was in possession of a knife and had threatened passengers on public transit, which initiated a police response. Second, Yatim's death was a result of a police shooting. In absence of social media materials, in particular a bystander-recorded video of the shooting uploaded to YouTube, information provided about the shooting to the public would have drawn from official accounts generated by police and presented in news media.

Social media platforms allow social scientists interested in examining definitional meaning making to peer into the communication process as it unfolds and observe when and how meanings are produced and negotiated. This approach can provide important insight into developments in social order. "Social order is about shared definitions, including temporal and spatial configurations for realizing and enacting definitions" (Altheide and Snow 1991, 4). The video of Yatim's death generated hundreds of print, radio, and television news media reports and thousands of online comments. The sum of these materials contributes to the collective meaning-making process of the death, including how the death becomes framed, defined, and understood. Moreover, the "high-profile streetcar shooting of Sammy Yatim in Toronto represents an extreme example in a growing number of cases where citizen surveillance videos are used by oversight agencies for their investigations" (Campeau 2015, 2). What makes this circumstance extraordinary is that the recording of Yatim's death appeared on social media in advance of news media or police accounts. Because of the widespread proliferation of communication and information devices, such circumstances are only likely to increase in frequency. This chapter chronicles the beginning stages of this circumstance—one where police initially exhibit far less control over meaning making because of user-generated material on social media.

THE DEATH OF "ARMED VICTIM" SAMMY YATIM

Eighteen-year-old Sammy Yatim was shot to death by Toronto Police Constable James Forcillo in the early morning of July 27, 2013. Shortly after, a ninety-second video recording of the shooting, taken by Martin Baron, was uploaded to YouTube. Much speculation has surrounded the circumstances of the shooting—a great degree of which draws from various recordings of the incident. Future details will no doubt continue to emerge as the trial of Constable Forcillo began in November 2015, as this book went to press.[2]

Here is a chronology of the relevant details: First, Yatim brandished a knife aboard a Toronto public transit streetcar. When police arrived at the scene, all passengers had vacated the streetcar, including the driver. Yatim was ordered by responding officers to "drop the knife" several times before nine shots were fired (eight of which reportedly struck Yatim). Finally, Yatim was tasered immediately following the shooting.[3] The officer that fired all shots was Constable James Forcillo, and a second unnamed officer tasered Yatim. Three weeks after the shooting, Forcillo was charged with second-degree murder (*Toronto Star* 2013b). A year later, in July 2014, the Yatim family filed a multimillion-dollar lawsuit against the Toronto Police Service for "reckless disregard" for the life of Sammy Yatim. At the end of July 2014, and almost one year to the day of the shooting, a second charge of attempted murder was laid by the Crown against Constable Forcillo. Much speculation ensued in the press about the rationale for this second charge following Forcillo's preliminary inquiry in June 2014; evidence was not reported due to a publication ban. Details of the shooting will certainly continue to emerge. My concern here is to better understand how Yatim's death came to be defined in relation to a recording of the shooting that was uploaded to YouTube as well as other material, including subsequent recordings and user-generated comments, which also contributed to the construction of news media narratives of the shooting.

Consider the headline of a July 30, 2013 *National Post* report: "Police Fired Multiple Shots at *Armed Victim*" (Boesveld 2013, emphasis mine). This article presented Yatim as a victim of police, a dominant theme in news media reports. However, it is important to stress that victims are very rarely described as armed. What social conditions make this headline and other similar statements possible? The response to this question, I wish to suggest, can be best understood through an investigation of the sense-making process on social media immediately following the Yatim shooting. I argue that how Sammy Yatim was framed on social media contributed directly to the meanings associated with the shooting, including the advancement of the "armed victim" storyline.

NEWS MEDIA REPORTS OF CITIZEN DEATHS DURING POLICE ACTIONS

It is impossible to know for sure how much traction the shooting death of Yatim would have gained in news media reports in absence of the citizen-recorded YouTube video. While forms of police use of force can be left out of news media or managed internally by police (or both), it is near certain that Yatim's death would have received some coverage. A *Toronto Star* report, paraphrasing lawyer Peter Rosenthal, noted that what makes the "case unique among similar cases involving police killings is the graphic video evidence taken by bystanders and posted to YouTube" (Hasham 2013). Reports of civilian deaths at the hands of the police in the days following the Yatim shooting support the strong likelihood of news coverage even in the absence of video. A few examples illustrate the point.

It was reported that two men on two separate occasions, on August 2 and 3, 2013, respectively, had died as a result of police actions in Alberta.

On August 2, 2013, news media reported that a man had died in Alberta one day following the police use of a taser weapon on him. The Alberta Serious Incident Response Team (ASIR) has been called to investigate these deaths. No public recording of these deaths is known to exist. Perhaps for this reason, these deaths have received minimal national news coverage compared to the Yatim shooting. In the Alberta incidents, news media relied on official information provided by police, which is consistent with the aforementioned research that identifies police as the primary sources of crime-related information. A *Globe and Mail* article reporting on the Alberta deaths illustrates the point. Regarding the use of a taser on August 2, 2013, "*RCMP said* the man who they used a taser on during the arrest Friday was a suspect in a series of assaults, theft of automobiles, driving complaints and hit-and-runs. They said they stopped him at a gas station, but they said there was a physical confrontation and that's when the taser was used" (Canadian Press 2013, emphasis mine). Regarding a police shooting on August 3, 2013, "*RCMP have released few details about the shooting, but they have said* two officers were trying to pull over a suspected impaired driver when a confrontation ensued and a man was shot" (Canadian Press 2013, emphasis mine). The report of these civilian deaths, framed in terms of confrontation resulting in the police use of lethal force, also stressed the significance of the Yatim video footage:

> The debate over how much force police should use was propelled into the national spotlight last week after the fatal shooting of Sammy Yatim, 18, in Toronto last month. Yatim died after being shot by police aboard an empty Toronto streetcar he had cleared out by brandishing a knife. Video of the incident has been viewed by many people online and his death has sparked protests and prompted an online petition calling for criminal charges to be filed against the officer who fired the shots. (Canadian Press 2013)

While civilian deaths as a direct result of police use of force in the Western world are relatively infrequent, civilian recordings of these incidents are certain to increase, due in large part to the pervasiveness of recording devices. As social media platforms continue to develop and expand, empirical work remains necessary to further clarify connections between format, meaning, and discourse in relation to contemporary understandings of police use of force. This now includes the growing inability of police to control definitions by providing journalists with narratives that defend the police use of force (Lawrence 2000).

THE SAMMY YATIM VIDEO RECORDING

The citizen-recorded video that documented the circumstances that led to Sammy Yatim's death (it was reported that he died at the hospital) in direct relation to police action is just one of a few citizen recordings in Canada to capture a police shooting that resulted in death.[4] It is not, however, the first citizen-recorded video in Canada of police action resulting in death. The most widely documented was the 2007 video of Polish immigrant Robert Dziekanski being tasered by RCMP officers at the Vancouver International Airport. The British Columbia Corners Service ruled the death a homicide in 2013. The Dziekanski and Yatim videos both received extensive national news coverage. The Dziekanski video, which was initially seized by police at the scene and later released to news media by Paul Pritchard, the bystander who recorded the video, showed footage that directly contradicted official statements of the circumstances given by the responding RCMP officers (Braidwood 2010). The video ignited a national firestorm of criticism directed against the RCMP. As Andrew Goldsmith suggests, the Dziekanski footage was "key . . . to the public accounting of this incident [and] while it is indeed one account only of what occurred, its visual form provided a potency and status in the ensuing investigations and public discussions that far outweighed other sources and points of reference relating to the airport incident" (2010, 925).[5] The Yatim video, on the other hand, was given no context in news media or by police in advance of the "release" (i.e., uploading) of the video to YouTube. This is a rather significant distinction in terms of how the video was first framed online and quickly thereafter in news media reports.

The ninety-second Yatim video fast became the primary source for news media. In fact, for a very short time, it was the only source. Other sources emerged later, including alternative angles and CCTV footage as well as video recording with enhanced audio. The sourcing of the video and other materials from social media provides content for user-generated definitions of the situation that helps contribute to the framing process and resultant discourse about the shooting.

INITIAL RESPONSE ON YOUTUBE

News media sourcing of the recording also drew increased attention to the video on YouTube, as evidenced in examined user posts that referenced news media. Baron uploaded the original video recording of the Sammy Yatim shooting on YouTube on July 26, 2013.[6] This version was viewed 612,825 times in one week, and a total of 8,586 user comments were made. The video was hosted on various websites alongside alternate versions that later emerged. According to the *Toronto Star*, the video had been viewed "about one million times" around the world in the few days that followed the shooting (McDiarmid 2013). By August 20, 2013, the CBC reported the total number of YouTube videos that included several versions of the original video were collectively viewed "1.8 million times and probably several times that number have seen the video on television or on other websites" (Schwartz 2013). In the eight weeks that followed, the original Baron recorded version received an additional 27,185 views and only 267 user comments—a decrease of nearly 97 percent in user postings. For this reason, those comments made in the week following the shooting provide the most suitable time frame of analysis and insight into how the event was initially framed online.

The Yatim video emerged online in advance of frames presented in news media and by police of the shooting—a very rare circumstance, especially considering the situation of a citizen death as a result of police action. The following posts made shortly after the video was uploaded to YouTube help illustrate the point. One commenter asked, "Was this on the news?" Another asked, "What will the cops say?"[7] A review of comments posted to YouTube provides some early perspective into the development of online collective user interpretation. But, more importantly perhaps, these comments help us to better understanding the relevance of YouTube as a communication format that contributes to the discourse associated with the construction of the Yatim shooting. The YouTube video provides various audiences simultaneously—users, news media, and police—with an early and basic frame of reference, one that unfolds parallel to news media and police investigation. This space on social media modifies institutional practices, on the one hand, and helps reveal the way that messages on YouTube account for how meaning is constructed, on the other hand.

FRAMING THE SHOOTING

How was the Sammy Yatim video framed on YouTube? And in what ways did this framing contribute to the social construction of the shooting? How the video is framed, first in advance of official claims-makers, helps provide a basic focus and context for the ways in which online users might interpret the video. This process contributes directly to the emergence of user-generated themes expressed in the form of user narratives that include posts presented in response to video.

Baron, who happened upon the scene of the police confrontation with Sammy Yatim on Dundas Street West near Bellwoods Avenue, video recorded the altercation on his iPhone. He then uploaded the video to YouTube. Baron later also posted a tweet embedded with the video at 1:23 a.m. EST with the following text: "To witness a shooting is horrible, but in front of your house?" The text description provided by Baron that accompanied the video on YouTube and the tweet read, "18 year old Sammy Yatim was holding up a knife on an empty streetcar. The cops surrounded him and ordered him repeatedly to drop it. He wouldn't, then they shot him. Listen for the Taser at 1:01." This post was the first published report that Yatim was in possession of a knife *as a weapon*. Police commands to drop the knife provide this context, whereas various news media reports contextualized Yatim primarily as a knife enthusiast. A *Globe and Mail* report in the days following the shooting also suggests that Yatim was a knife enthusiast, ahead of an explanation of the knife as a possible weapon, a dominant narrative that then developed as a part of the post-shooting narrative in news media:

> It is not clear exactly what happened in the moments leading up to the shooting on Saturday just after midnight. Witnesses say that Mr. Yatim, a knife collector, brandished a knife on a Dundas streetcar at Bellwoods Avenue and exposed himself. Everybody rushed out of the streetcar leaving him behind on the vehicle. In a witness video of the police surrounding the streetcar, nine shots are fired after which police can be seen and heard going in to taser Mr. Yatim. (Kauri 2013)

The YouTube and this *Globe and Mail* description provide a similar context (the empty streetcar), but they frame Yatim's possession of the knife differently. The social context presented in the YouTube description, which was reiterated in the *Globe and Mail* article and in the embedded original video in the online version of the article, provides audience members with an interpretative frame that directs viewers what to look and listen for when viewing the video, thereby influencing the parameters of the ensuing online discourse. The importance of the video, along with the video description, is demonstrated in YouTube–facilitated discussions and news media reports that illustrate the knife as a key topic of focus. Eyewitness accounts in news media of the shooting confirmed that Yatim had a knife, in support of statements provided by amateur videographer Baron himself. A *Toronto Star* report published on July 28, 2013 draws from Baron's own recollection of the shooting, which also appeared in other news media reports:

> Police were not confirming Yatim's identity Saturday night, but friends and acquaintances were in shock and mourning in conversations on social media. When passersby Heather and Martin Baron first saw a stalled streetcar, they thought it was broken down. But as the pair got closer, they saw about five officers swarming the front door of the streetcar, yelling at a knife-wielding man beside the driver's seat, "Drop the knife! Drop the knife!" (*Toronto Star* 2013a)

This report brings to our attention that police were not in control of the publication of Yatim's name online, let alone the video, but, more importantly, that the framing of the event can be directly linked to Baron's video description and recollection of the shooting. Baron's video is recorded from a distance, reportedly thirty-two feet away, and it is difficult to see if Yatim is in possession of a knife. One user post on YouTube read, "Anyway, I don't see the knife."

Other versions of the video quickly surfaced, including another iPhone video recorded by bystander Markus Grupps. This video was reported in news media as an "enhanced audio" version where Yatim could be heard telling an officer "you're a fucking pussy." The same day, on July 29, the *National Post* published a report with the headline "Sammy Yatim's Final Warning: New Audio Reveals Officer's Hostile Words before Teen Was Shot to Death by Police." The online version of the article, embedded with an audio clip, reported that the "enhanced audio recording . . . was filtered from video of the incident posted on YouTube" and also features comments from Baron (Visser 2013).

The description on YouTube provided an early context and frame of the shooting that developed as the dominant frame. This set the perspective and tone of the police shooting and provided a powerful influence over taken-for-granted viewer assumptions of what had happened, on the one hand, and the moral discourse that quickly ensued online and in news media reports, on the other hand. It is important to underscore that the belief in the presence of a weapon in the altercation with Yatim plays a significant role in how the meanings surrounding the shooting become interpreted by users as justified or not. Indeed, the knife became the primary focal point of interpretations of the shooting by users and in news media reports.

DISCUSSION OF THE KNIFE

The word "knife" appeared no less than 2,100 times across user postings in a single week on YouTube where the original Baron video first appeared. While other related words (e.g., "weapon" and "blade") appeared hundreds of times, these terms were less frequent. Explicit words such as "threat" and "danger" sometimes appeared alongside user discussions of the knife. However, the word "knife" alone, absent of an adjective (e.g., kitchen knife or hunting knife), given the confrontation between Yatim and police, provides at least the suggestion of danger.

In the days that followed the shooting, the knife developed as the key topic. The knife was described in news media as "small," "little," or "three inches" and, on one occasion in the *Toronto Star*, as "itty-bitty." In news media, Yatim was near universally described as "brandishing" or "wielding" a knife (the two most frequent terms used) and intermittently as either "carrying" or "holding" a knife. In a very few circumstances in news media reports, Yatim was characterized as "armed" with a knife. One of these reports was an editorial by noted controversial columnist Christie Blatchford. It is important to stress here what was not discussed. In news media, Yatim was not reported as "threatening" with a knife. This was not the case on YouTube, where a multitude of users discussed

Yatim as threatening. One user post the day after the uploading of the video read, "the guy had a knife and he was threatening people with it."

A range of user interpretations about the knife were presented on YouTube. The knife was variously described as a "pocket knife," "butter knife," "2 inch knife," "3 inch knife," "kitchen knife," "large knife," "K-bar" (likely a reference to the Ka-Bar brand military combat knife), and, in one circumstance, "sword." While discussion about the type of knife ensued across user posts, it is important to note that Yatim's possession of a knife was not at all contested, nor were police commands. Furthermore, these circumstances were also not contested in news media reports. As one user remarked, "There is no question that the police asked him to drop the knife." In fact, the users interpreted the presence of a knife as a distinct threat that necessitated police use of force. Because of the knife and Yatim's failure to comply with police commands, users did not interpret police use of force as police brutality: "this is not police brutality people, this is a police officer neutralizing someone who was too dumb to put down the knife."

The failure to comply with police commands as heard on the video allowed a lot of users to perceive and interpret Yatim as deserving of his fate: "Perhaps he shouldn't have wielded a knife at police. If you catch me ding that, shoot me"; "kid got what he deserved, only morons pull out knifes on cops." Other users expressed disapproval that the situation could not have been resolved by otherwise peaceful means. The knife and police commands framed these and similarly offered responses, including, for instance, how users imagined themselves in the situation as an officer. A user who self-identified as a police officer wrote, "I come from a third world country. I was a cop back then and as far as I know you don't need to kill a person with a knife. You are supposed to disabled him by shooting to his legs or hands." Some found imagined scenarios an agreeable a posteriori response to the shooting: "If I was in that officer's situation, I can easily think of what I would have done. I would have shot him in the legs. It's an easy shot for someone with only limited experience, and in this hypothetical situation I have police-grade firearms training." Many users, however, dismissed these imagined scenarios as unrealistic perspectives associated with entertainment media: "shot in the arms or legs? Wow clearly you watch too many movies." Others noted that this type of officer response was not in accordance with "real" police training procedures. YouTube users pointed others to sources to support their ideas. One commenter wrote, "look up this video titled: 'Too close— don't underestimate the knife,'" a three-minute clip from the 1988 VHS police training video *Surviving Edged Weapons: How to Defeat Today's Fasted Growing Threat*. The video illustrates various staged scenarios that depict knife-wielding perpetrators attacking and stabbing at uniformed police officers. The voiceover narration provides some context:

With a reactionary gap of about one foot or less it's impossible for you to react quickly enough for you to even touch your holstered sidearm once that attack begins. At about five feet the average officer can't event get his sidearm unholstered. At about ten feet you might get your sidearm out but you probably won't get a shot off.

In line with this narrated training scenario, another user on YouTube asked, "Have you ever tried to shoot a crazed knife wielding person charging at you under stress? Police are never training to shoot arms or legs although it works well in Hollywood, real police shoot for the largest part of the body called centre mass, police will shoot until there is no longer a threat."

DISCUSSION OF POLICE USE OF FORCE

A basic theme of user posts involved police use of force and whether or not the force was excessive or justified. Dominant narratives discussed Yatim as either a helpless victim with a mental illness (excessive force, not justified) or as a knife-wielding perpetrator (justified use of force). Mental illness emerged on YouTube as a likely explanation of Yatim's behavior.[8] These accounts discussed and sourced news media reports that indicated that Yatim had exposed himself to passengers on the streetcar prior to the arrival of police officers. News media also reported that the Yatim family denied Sammy had a mental illness. These *vocabularies of motive* are words expressed in response to social actions that "stand for anticipated situational consequences of questioned conduct [where] a satisfactory or adequate motive is one that satisfies the questioners of an act" (Mills 1940, 905, 907).

The expressed belief on YouTube that Yatim was in possession of a knife for no apparent or otherwise socially sanctioned reason did not serve as a universally acceptable justification for police use of force, even while very few users accepted the knife collector account. There must have been a reason or motive for Yatim to be in possession of the knife in the first place, such as mental illness or the intent to harm. Context of possession and discussions of the knife and motive circulated on YouTube mostly in advance of police and news media statements offered in response to the video. There were a few exceptions. Yatim was reported as a knife collector in news media as early as July 29 (Kauri 2013). While we cannot be certain if YouTube users were knowledgeable of reported news media accounts, a post made the following day on July 30 read, "he was a knife collector and knife enthusiast. He was likely very dangerous and skilled with the knife. Very few officers are able to take on a knife expert in hand-to-hand combat." What this post and others like it seem to suggest is that user-generated discourse that attempted to make sense of the definition of the situation did so as this process developed alongside news media. In these circumstances, attention shifted to the motives of individual actors: the possession of the knife, on the one hand, and the amount of police force, on the other hand. Regarding police use of force, YouTube users focused heavily on the quantifiable (nine gunshots) amount of force. Yatim was framed as a perpetrator who was up to no good, mentally disturbed, and often both. One user wrote, "We live in Canada man why have a knife? Unless you are going to kill someone." Another noted, "If a person poses a threat by waving a knife, & approaching police, I don't care if their off their meds, autistic or diabetic. Drop them." The issue of mental illness was also a recurring and basic user-generated theme. Some users referred to Yatim as "bipolar" or "mentally disturbed," whereas others noted that he was a "deranged" or "textbook psychopath" that was "having a breakdown." Drugs or alcohol were also provided as accounts to either explain or excuse the police shooting, as was "suicide by cop."

YouTube commenters discussed the police use of force variously as between justified and excessive: "Sammy was just plainly murdered! 9 shots overkill and tasering a dying boy as torture." Force was also conflated as both justified *and* excessive, as one user noted: "I think the first three shots would've been more than enough." Many others expressed confusion that the taser was deployed *after* the shots were fired. It was a shared belief by most that the taser use was acceptable and should have either been the only force used, or, at the very least, have been used before the shots were fired. Because others believed that the police weapons and Yatim's "weapon" were mismatched, the force exhibited by police was viewed as inappropriate. As user one put it: "A knife versus a gun . . . the kid was basically unarmed! Unnecessary excessive use of force."

OFFICIAL POLICE RESPONSE ON SOCIAL MEDIA

Given the TPS's extensive use of social media, as detailed in the previous chapter, it is particularly striking how little information was actually provided on social media and in news media in direct response to the shooting. Legal reasons were provided as one justification; however, this explanation was not enough to thwart criticism of the police. The documentation and near instantaneous circulation of various discrediting materials on social media platforms problematizes police "account ability" (Ericson 1995) and can make police "more vulnerable and exposed" than ever before in certain circumstances (Goldsmith 2010, 920). This "new visibility" (Thompson 2005) provides new opportunities for the public to scrutinize police on social media and greatly expands the parameters of police "account ability" (Ericson 1995).

As Richard Ericson explains, "Accountability entails an obligation to give an account of activities within one's ambit of responsibility" (1995, 136), whereas "account ability [refers to] the capacity to provide a record of activities that explains them in a credible manner so that they appear to satisfy the rights and obligations of accountability" (137). Account ability as a police strategy of negotiation relies on practices of secrecy, on the one hand, and revelation, on the other hand, each operating in tandem (Ericson 1995; see also Goldsmith 2010). However, in cases such as the Yatim shooting, the TPS were less able to practice secrecy and were not able to control the *immediate* revelation of details related to the shooting—details ascertained by the public and news media from the circulation of the video recording of the shooting on YouTube. Social media generate interactive environments for the display of "account ability" from which police can explain their activities to members of the public. This includes "followers" on sites like Twitter where the TPS are the most active in terms of their social media strategy. For instance, in the two days that followed the shooting of Yatim, @TorontoPolice posted fourteen tweets about the shooting. At the time, the @TorontoPolice feed had 50,552 followers.

The law was directly invoked in two TPS tweets and indirectly in three other tweets that noted the "ongoing investigation" of the shooting. A tweet posted on July 30 on @TorontoPolice read, "Sammy Yatim Shooting: By law Toronto Police cannot comment further than Chief Blair's statement July29 –Watch video." Deputy Chief Peter Sloly also tweeted: "Appreciate sentiment of all who engaged me in last 24hrs—can't comment further re SIU but offer thoughts/prayers to all directly affected" (July 30, 5:00 a.m.). Several other officers, including Constable Scott Mills, retweeted versions of these tweets and some officers retweeted supportive tweets from the public. For instance, on July 29 at 11:10 a.m., Staff Sergeant Chris Boddy retweeted the following from a member of the public: "@TPSChrisBoddy as a former mayoral candidate I support the police in their actions—but this is so incredibly disappointing. #SammyYatim."

Invoking the law in this manner is consistent with police account ability in news appearances (Ericson 1995). Social media also provide new opportunities for police to retweet supportive statements from members of the public. A link in the above-mentioned @TorontoPolice tweet, which was retweeted by several officers as well as members of the public, directed users to view an official video statement by Chief William Blair. Blair's statement was presented to news media the morning following the shooting. In advance of the statement, @TorontoPolice tweeted at 8:09 a.m., "Chief Blair is about to address the media from the Gallery at HQ regarding Friday night's shooting" (July 29, 2013).[9] The statement was formally presented to news media, and a version was posted to the official TPS YouTube channel, which demonstrates that social media platforms now operate alongside news media as a key feature of the accountability structure in criminal justice (Ericson 1995).

In the video statement posted to YouTube, the same platform where the Yatim shooting video surfaced just hours earlier, Chief Blair explains,

> The Special Investigations Unit has invoked their mandate and their investigation must take priority over all other inquires. I am prevented by law from disclosing any information with respect to the incident, or the investigation. This regulation is aimed at maintaining the integrity of the investigation and I intend to uphold it . . . addressing the concerns of the deceased young man's family is our highest priority. We will act as quickly as circumstances of the law allow. (TPS 2013b)

Sanctioned official police posts offered in response to the Yatim shooting were not located on YouTube, where the original video was featured, short of a single related post made by the Toronto Police Association (TPA), an organization that officially represents the interests of police. The TPA is one of the largest organizations of its kind in North America. Its mandate is to advocate on behalf of the police profession to influence internal and external decisions that affect its more than eight thousand uniformed and civilian members. This includes lobbying efforts directed at government and politicians as well as promoting the integrity and value of policing among the general public. These outreach efforts occurred on YouTube in response to the Yatim shooting. The single TPA post offered in response read, "It is important for everyone to wait for facts & evidence of SIU [Special Investigations Unit] Ontario." Many users on YouTube were quite skeptical of the SIU. One asked, "Wait for SIU investigation why? Video is clear enough for me! Shooting a boy 9 times while cops walk around bored, then tasering Sammy while he was dying. What evidence can possible change these facts?" Other users questioned the integrity of the SIU: "And you know the SIU will clear these fucks. . . . Thats what happens when cops investigate cops . . . what a joke." The mandate of the SIU is "to maintain confidence in Ontario's police services by assuring the public that police actions resulting in serious injury or death are subjected to rigorous, independent investigations" (SIU 2015). The independence of the SIU was an issue for some users and perhaps for good reason. The SIU was created by Ontario's Police Services Act Section 113, which states,

> (3) A person who is a police officer or former police officer shall not be appointed as director, and persons who are police officers shall not be appointed as investigators.
> (4) The director and investigators are peace officers.

According to the *Criminal Code of Canada*, a "peace officer" includes:

> (a) a mayor, warden, reeve, sheriff, deputy sheriff, sheriff's officer and justice of the peace, (c) a police officer, police constable, bailiff, con-

stable, or other person employed for the preservation and maintenance of the public peace or for the service or execution of civil process.

The TPA YouTube account from which the above post was made was created on July 29, 2013, less than three days following the Yatim shooting, which users readily observed on YouTube. One post read, "Come to youtube with your police propaganda. You created this account with the purpose of defending this heinous crime, which is why it was created today. gtfo [get the fuck off] youtube." According to its webpage, the "fundamental purpose" of the TPA is to "protect those who protect others." The above-mentioned TPA post made to YouTube also appeared on official TPA Twitter and Facebook accounts, launched June 8, 2009 and December 10, 2010, respectively. TPA used these social media platforms to counter numerous user claims advanced online but did not engage directly with any users online. The TPA also used social media to stem the tide of growing criticism directed against the police, especially by important claims-making organizations such as the Ontario Federation of Labour (OFL). The TPA published a press release on Facebook and links to this release on Twitter in direct response to statements made by the OFL. In the days that followed the Yatim shooting, the OFL published an official statement that called the shooting a "total failure of the Toronto policing system" (OFL 2013).

The July 31, 2013 OFL online statement noted, among other things, that "by all accounts [Yatim], did not pose any immediate threat to officers or bystanders" (OFL 2013). The OFL press release also references the Baron video of the shooting. Irwin Nanda, OFL executive vice-president, wrote, "Thanks to the graphic video evidence available, we can see the events unfold, but we are left with many questions about why this incident resulted in tragedy." Sid Ryan, OFL president, also referred to the shooting as "dangerous consequences of police over-reactions to mental illness." The official response to these statements by the TPA published on August 1, 2013 on Facebook calls these statements "uninformed and irresponsible" and states that OFL President Sid Ryan "has no knowledge of the vast amount of evidence being currently gathered and reviewed." The TPA statement continues:

> Mr. Ryan and the OFL know full well that the Special Investigations Unit immediately launched an independent investigation which is still on-going. Mr. Ryan has never been a police officer and has no expertise in policing or anything related to policing. . . . Mr. Ryan and the OFL suggest that Sammy Yatim had a mental illness. This is certainly news to the Toronto Police Association. While the TPA leaves open the possibility that mental illness may have been at play we have been given no confirmatory evidence. We are waiting for the results of the full investigation of all the evidence concerning this incident.

Immediately following the shooting, the SIU posted a single tweet that confirmed that it was investigating the incident. At the time of this writing, the SIU was only on a single social media site—Twitter, having joined on January 17, 2013. The SIU tweet was retweeted by various TPS officers. The TPS (@TorontoPolice) also tweeted, "Toronto Police are prohibited by law from commenting on officer involved shooting. Inquiries + witnesses call 1-800-787-8529" (July 27, 2013, 5:42 a.m.). TPS officers retweeted this tweet as well as links to the authorized TPS YouTube channel. These links directed to the official statement made on YouTube by TPS Chief Blair as noted above. The police response in *direct* relation to the Yatim shooting on social media was virtually non-existent and minimal at best. The framing of Yatim's death, especially in the early hours and days following the shooting, emerged largely from online materials and mostly in absence of police statements provided on social media.

SOURCING OF SOCIAL MEDIA MATERIALS: CONTESTS OF CHARACTER

Jonathan Goldsbie, a staff writer at Toronto's *NOW Magazine*, tweeted links to two screenshots of Sammy Yatim's Facebook profile, each taken on Sunday July 28, just one day following his death. The tweet read, "Sammy Yatim's Facebook profile was posthumously modified" (July 29, 2013, 7:21 a.m.). According to a subsequent tweet authored by Goldsbie in response to queries from another Twitter user, "Those are my own screenshots. The one with the shotguns was from Sunday afternoon, and the other is from late that night" (July 30, 2013, 7:30 a.m.). Links to these screenshots also appeared on the YouTube page where the original Baron video of the shooting was uploaded. It is not known for certain who modified Yatim's Facebook profile. One YouTube user wrote, "the kid appeared to be a thug by his posts on facebook, which have been censored by his parents in an effort to make him [not] look thugish."

The original screenshot posted by Goldsbie is of the top portion or banner picture of what appears to be Yatim's personal Facebook page. The image depicts no less than five AK-47 assault rifles in a circle (Goldsbie incorrectly identified these weapons as shotguns). The original and posthumous modified profile each indicates that Yatim was from Aleppo, Syria. The unmodified profile shows a picture of Yatim holding a bottle in his right hand. Evidence suggests that news media likely viewed the original profile *before it was modified*. A July 28, 2013 *Global News* television report broadcast the bottle image with the following voiceover narrative: "He lived in Toronto and was from Aleppo, Syria according to his Facebook profile."

It is important to note that the Goldsbie screenshots do not reveal the entire contents of either the original or modified Facebook profile page.

In the tweet of the original profile, the screenshot shows three partial small images on the left and one partial larger image on the right. Sixteen response tweets follow. The first of these tweets that followed nine minutes after Goldsbie's original tweet of the screenshots read, "Saw that on Reddit. I don't think this sort of Facebook rummaging ever helps anyone's understanding of tragic events" (July 29, 2013, 7:30 a.m.). Reddit is akin to a social networking site but operates as an online forum bulletin board where users post links and comments. Subforums or "subreddits" allow users to locate and follow specific topics of interest. Reddit posts can be made in a more anonymous manner than on YouTube or Facebook, where users must have an account to post.

Links to the image-hosting site Imugr directed online users to more complete screenshots of Yatim's original unmodified Facebook profile. These screenshots appeared under the subreddit thread "mentalhealth-problems." These images mostly show Yatim with friends. In some of these photos, however, Yatim is shown brandishing two bottles of alcohol, and is seen holding cigarettes, and his friends are holding cigars. In other images, Yatim is shown alongside young men who collectively appear to be flashing gang signs and finger guns pointed in the direction of the photographer. Links to these images on Reddit were accompanied with user interpretations such as, "The real guy Sammy Yatim who got killed" and "here is who the police really shot." Other text posts referred to Yatim as a "goon," "tough guy douche bag," and an "extra ballsy problem starter." Another noted, "I don't think its necessarily mental illness, for then all his friends must be suffering as bad or worse with their tough guy bully mental problems." Other users were critical of the "rebranded" Facebook page that "whitewashed" his past, while some pointed out that "the kid doesn't seem to be an angel like his family is trying to argue."

These screenshots were viewed on Imgur more than nine thousand times and generated more than five hundred comments on at least one Reddit thread. Discussion and display of these images *were ignored almost entirely in mainstream media reports*, with just a single exception—a *Toronto Sun* report published August 2, 2013: "The truth is Sammy Yatim was not the angel some would like to make him out to be. He had a knife, had been reported to be involved with lewd behaviour prior to the incident [it was alleged that he had exposed his penis to others on the streetcar] and on his Facebook page is a picture of guns. But he was a kid and certainly did not have die" (Warmington 2013).

While Yatim's personal Facebook profile was largely ignored in mainstream media, news media did, however, source a great deal of materials from other Facebook pages created after his death, including one called "Sammy's Fight Back for Justice," which was one of the first to appear post-mortem (on July 28), launched by friends and the sister of the deceased, Sarah Yatim. The description of the Facebook page explained,

"Sammy never had a criminal record, mental illness, or suicidal tendencies." The knife was also a topic of focus. The more than eight-hundred-word description continued: "Sammy Yatim was carrying a pocket knife just because he enjoyed collecting them back at home. He brought it out of his pocket and was switching it out in his hand (similar to a Swiss Army knife that many carry). Word passed through the streetcar and they called the cops." Narratives emerged on this Facebook page about Sammy's life and his personal character. The first status update made by the moderator, someone called Anita, reportedly one of Sammy's friends, stressed the need for "awareness" to get out the "RIGHT" story, the truth that Sammy was "innocent" and "shot to death for nothing." In fact, multiple accounts were made on "Sammy's Fight Back for Justice" by various people that self-identified as his friends or family. One post read, "I am really upset and all I can think about is my 18 year old cousin Sammy, killed for no good reason. He was a good guy and we all knew" (July 29, 2013, 12:02 a.m.). Another read,

> Just about a month ago Sammy was sitting at my home smoking shisha with me. He went to my high school and was a good friend of mine. Honest to God, he was a straight up nice guy, never had any intention to hurt anyone and always cared about his boys. I can't believe this happened to him, and in the city of Toronto. The level of brutality and the fascist nature of Toronto police has shocked all of us. They disregard human life and killed my friend—a 18 years old boy who was just about to start his life. I want justice, and PUNISHMENT to be brought to Sammy's killers. (July 28, 2013, 5:12 p.m.)

These posts provided meaningful accounts for others: "I did not know Sammy [but] from what I hear and see, he was a sweet kid (July 29, 2013, 11:31 p.m.). Another post read, "all witnesses say he wasn't attacking anyone, and his family and friends say he would never attack a police with a weapon, he was a good kid, going to school and whatever, you cant assume he was some criminal thug" (July 28, 2013, 8:49 p.m.). Anita indicated that she would "post pictures [on the page] to use on the posters [during a protest that took place that evening in Toronto] to portray Sammy's true character!" One of the poster images and also the Facebook profile picture of the group page, and therefore possibly signifying increased importance, was posted to encourage audience members and protestors to portray Yatim's "true character." The image, a portrait-style picture of Yatim facing forward in a baseball cap, appears have first surfaced on July 28 on "Sammy's Fight Back for Justice" but was credited in news media as being sourced from another Facebook page—the "RIP Sammy Yatim" page, created on July 29. The portrait-style, "boy next door" image was prominently featured in news media print and television reports, including CTV News, CBC TV News, CBC News online, and in national publications such as the *Globe and Mail* and the *National Post,* as well as regional newspapers, the *Ottawa Citizen*, and the Vancouver *Province*.

"Sammy's Fight Back for Justice" Facebook page was also featured in news media reports that covered the first of two protests in response to the shooting death of Yatim. For instance, an online CBC report titled "Streetcar Shooting Protest March Draws Hundreds" included an embedded link that directed users to the Facebook page. A disclaimer in the report indicated that the CBC does not endorse external links. The July 29 protest, which received favorable coverage in news media and was framed in terms of social justice, was promoted on "Sammy's Fight Back For Justice" by the moderator Anita like a rock 'n' roll concert: "Thanks for all the support everyone! Don't forget protest is today at 5PM at Dundas Square! (Todayyyyyyy July 29th, 2013). We have posters, a stage, a petition to be singed [*sic*], videos being played, t-shirts! Its going to be unreal! Everyone make sure you come!" (7:44 a.m.). Approved poster slogans were provided for protestors in advance, including "help us fight for the innocent killed by 'authorities'" and "protect us from our protectors"—a play on the TPA mandate. When suggestions were made about the possibility of the use of alternative slogans, Anita responded, "There is a reason we have asked people to NOT make their own slogans so please abide by the rules we have set" (July 29, 2013, 9:16 a.m.). Whatever the reason, none was provided.

Links to news media articles that reported on the Yatim shooting were posted by the moderators of the two most widely reported and popular Facebook pages "Sammy's Fight Back for Justice" and "R.I.P. Sammy Yatim." The first status update made to the latter page included a link to a *Toronto Star* report to help inform users about the shooting, which suggests tacit approval of news media accounts. The news report included statements provided by family members as well as a Yatim family friend. The friend discounted any speculation of untoward motives associated with Yatim's possession of the knife. A handful of the "R.I.P. Sammy Yatim" Facebook page status updates made by the moderator included endorsements in the form of references and news media links. An August 25, 2013 update illustrates the point: "Who was Sammy Yatim? Read about it here." The link directed users to a *Toronto Star* report that detailed Yatim's life, beginning with his "childhood" noting that he was born "in a middle-class Christian family." Close friends and family are sourced. So was a former teacher, Megan Douglas, who "fondly remembers Yatim as the polite boy with a stunning smile" (Alamenciak and Ghafour 2013). News media reports largely provided narrative accounts that, much like above-referenced Facebook materials, profile Sammy's life and his personal character, which together suggested that he was most undeserving of his fate.

SUMMARY

Crime stories are an important feature of news media (Chermak 1995). Previous research has shown that police exert *substantial* control over crime stories and accounts (Doyle 2006). This includes the early stages of crime stories. Traditionally, only minimal information about crimes was available, and journalists typically relied almost exclusively on police to provide data and facts. Communication and information technologies, coupled with social media, have irrevocably altered this process. Crime events can be happened upon or discovered in progress and documented by civilian bystanders who in turn may circulate these materials online. In the case of the Yatim shooting, these materials can become an early and primary source for news media in advance of police accounts.

The Sammy Yatim shooting death represents a watershed social media moment for police in Canada, and, like the Rodney King video, was probably seen everywhere in the world. The production of narratives, including claims and counterclaims, about the shooting quickly emerged on various social media platforms. Police are less able to control and manage interpretations of the use of force in news media when crime materials circulate on social media platforms, and especially, as this chapter shows, when these materials displace police information as the primary sources of news media accounts. The Yatim shooting may be unusual because of the existence of the video, but what makes the shooting especially unique is the public availability of the video online immediately following the shooting. How the video was framed on YouTube directly contributed to the collective meaning-making process of the death, including how the death was quickly defined and understood. Because of the widespread proliferation of mobile technologies capable of recording everyday social life, and the multitude of social media platforms, instances of documentation like that of the Yatim shooting are certain to increase.

A central concern for police in these kinds of scenarios becomes modifying institutional practices to accommodate social media, on the one hand, while attempting to re-exert control over authorized circulation of accounts and materials, such as video recordings, on the other hand. This seems to suggest a necessary sea change in how police will respond to future incidents of this kind. The introduction of social media logic represents the next development of this phase, which not only influences police work (Brown 2013), but also contributes to the shaping of police practices and procedures. A *Globe and Mail* report published several months after the Yatim shooting titled "Toronto Police Mull Body-Worn Cameras For Officers" illustrates the point: "The Toronto Police Service is considering outfitting all uniformed officers with body-worn video cameras. . . . Depending on the style, the small camera can be mounted on a pair of glasses or onto an officer's uniform, and *documents events from the officer's point of view"* (Hui 2013, emphasis mine). A pilot project that outfitted one hundred Toronto police officers with body-worn cameras was launched in May 2015. These cameras provide police with an official recording of any given situation, thereby rendering any and all other recordings that might circulate online as unofficial. In the words of Toronto Police Staff Superintendent Tom Russell, "We believe that body-worn cameras are a valuable piece of technology that will provide an unbiased, accurate account of our interaction with the public" (Mehta 2015). These records will likely help police regain control over the crime narrative.

This chapter details how a citizen-recorded video uploaded to YouTube quickly following a crime event brought attention to the shooting but, importantly, framed the shooting death of Sammy Yatim and provided an interpretative context for the emergence of meanings associated with the death. As this chapter illustrates, police have much less control over a meaning-making process that now includes sourcing various materials from social media platforms. What we cannot account for is why selected information was sourced in news media reports while other online materials were ignored almost entirely.

NOTES

1. For a more contemporary discussion of "police riots," see Williams (2007, 177–96).

2. See, for example, Shannon Kari's *Globe and Mail* article, "Forcillo Testifies He Was 'Trained to Win' at Trial for Yatim Shooting Death" (November 27, 2015).

3. At the time of the shooting, frontline officers in Ontario were not authorized to carry tasers. It was announced one month to the day following the shooting that taser access in Ontario had been expanded to include all frontline officers (officials denied that the decision was influenced by the shooting).

4. Another example includes a 2007 cellphone recording of Vancouver Police Department Constable Lee Chipperfield shooting at Paul Boyd, a mentally ill man, nine times (including once in the head). It was reported that eight bullets struck Boyd. The video surfaced five years later in May 2012.

5. See also Milbrandt (2010, 128–29) for a brief discussion of public visibility and accountability in relation to the Dziekanski video.

6. The date stamp on YouTube indicates that the original video—that is, the one taken and uploaded by Martin Baron—reads "Published July 26, 2013." The shooting occurred just after midnight, however, which makes the official date of the shooting July 27. I have referred to July 26 as the date of video, but, following news media reports, I refer to July 27 as the date of the shooting. The original video, called "Police shoot 18 year old Sammy Yatim at Bellwoods and Dundas, Toronto," can be viewed at http://www.youtube.com/watch?v=PuItH2raahg.

7. YouTube comments, tweets, and Facebook posts have been transcribed exactly as they originally appeared. They have not been edited for proper grammar or spelling.

8. This link was even made in the *Canadian Medical Association Journal*, despite the fact that "Toronto Police Service spokesperson Mark Pugash was unable to provide further information or arrange an interview [following the shooting], however, saying

'all of these issues are under review' in the inquiry into the shooting of eighteen-year-old Sammy Yatim" (Glauser 2013, 1485).

9. The "Media Gallery" as it is sometimes referred to is the location of Corporate Communications, which, according to the TPS website, "is responsible for internal and external communication from the Toronto Police Service." See http://www.torontopolice.on.ca/corpcomm/.

Conclusion

Policing on Social Media

The basic argument advanced throughout this book is that police use of social media has altered institutional police practices in a manner consistent with the logic of social media platforms. My goal has been to show how these practices increasingly include police performance on social media. As noted at the outset, the three case studies indicate that there is no universal consensus about police use and control of social media. While, in some instances, police are able to effectively control and manage information, there are other circumstances where such command is problematic.

Another key point of the book is to show how attempts to make sense of events have become much quicker and more media-focused. My examination of this process reveals developments in the relations between social media and institutional police activity. The preceding case studies, however, are largely specific to two police departments in Canada and merely scratch the surface of developments in the area of policing and social media. Much work needs to be done to develop a comprehensive understanding of these contemporary police practices. Scholarship remains necessary to further understand how police agencies around the globe are responding to social media and how these new developments may have impacted and changed forms of policing and police work.

INTERNATIONAL DEVELOPMENTS

Police and law enforcement agencies will continue use social media, and evidence seems to suggest that these agencies will incorporate the logic of social media into policing and police work. I began this book with a discussion of media formats and social control practices. To paraphrase the quotation that began the first chapter, it is no longer a question of whether police will use social media but *how* police will use social media. I have developed media logic as a key analytic and qualitative media analysis as a methodological approach to the study of policing and police work. Public policing practices on social media will increasingly be governed by the rules and logics of mediation. I will briefly consider relevant examples from a few other countries to illustrate this point.

The United Kingdom Association of Chief Police Officers (2013) *Guidelines on the Safe Use of the Internet and Social Media by Police Officers and Police Staff* is one example of numerous law enforcement statements offered in response to the advent of social media. Section 4 of the document provides the following general guidelines:

4.1 The same standards of behaviour and conduct apply online as would be expected offline.

4.2 Information placed on the Internet or social media could potentially end up in the worldwide public domain and be seen or used by someone it was not intended for, even if it was intended to be "private" or is on a closed profile or group. It is likely that any information placed on the Internet or social media will be considered to be a public disclosure.

4.3 The public expect police forces, police officers and police staff to act with integrity and impartiality whilst upholding fundamental human rights and according equal respect to all persons. Police officers must abstain from any activity which is likely to interfere with the impartial discharge of their duty, or to give the impression to the public that it may interfere and must abstain from any active role in party politics.

4.4 Police officers and police staff should avoid using the Internet and social media off duty after consuming alcohol or when their judgment may be impaired for other reasons.

4.5 The use of social media for private purposes during working time and from force systems should be in accordance with local force policies. The use of social media for such purpose during working time, and from personal mobile devices, is not recommended.

4.6 Police officers and staff should be familiar with related force policies and procedures relating covering topics such as Information Security, the Data Protection Act 1998, and use of force Internet and e-mail.

While there are some similarities here to Canadian police social media policies, the activities of police officers in selected Canadian jurisdictions might not necessarily be entirely consistent with these UK guidelines. These differences may include use of social media while off duty, or perhaps even discussions on police social media accounts of officer allegiances to a particular hockey team. These activities may constitute use of social media for "private purposes" not related to police matters. One example might include a retweet posted by TPS 33 Division that, in part, read, "I had a couple of beers, so I'm on the streetcar!" (December 8, 2012, 9:57 p.m.). Tweets like this one would appear to be inconsistent with 4.4 of the UK police guidelines.

Public expectations that police operate on social media sites such as Twitter might also differ internationally, and so too might police use of these media to accommodate public expectations. One international example is Germany. As the *Babbage Science and Technology* blog on the *Economist* explained in a 2013 posting, "Germany ranks 31st worldwide in terms of public tweets, with 59m per year. Germany's 82m people have just 4m Twitter accounts. That puts it 22nd in the world, behind not only European neighbours like Britain (population 63m, 45m accounts) or Spain (population 47m, 16m accounts) but also Turkey (population 75m, 11m accounts) and the Philippines (population 98m, 8.6m accounts)" (S. W. 2013). This kind of information seems to suggest that the expansion and development of police efforts on Twitter might be *less* beneficial to advancing German police interests. Even still, evidence suggests that German police are using other social media such as Facebook to facilitate police social control efforts (Wünsch 2012) but that these efforts are ahead of German policy. A basic issue involves police control on social media, noted in chapter 1: "once something is posted, the content belongs to Facebook and can't be deleted. Unlike a poster on a wall, which can be removed, a Facebook post remains stored on a server" (Kern 2013).

Further research about variations among national and international policing institutions will reveal key differences in police practices online, social control strategies, and use of social media data. An international team of researchers might be best suited to further explore policing efforts on social media in other countries from a cross-cultural perspective.

MEDIA FORMATS AND CONTEMPORARY POLICE PRACTICES

I have argued that social media alters policing practices. Indeed, this book outlines how social media emerged and moved from outside the periphery of policing where it was first used as a "virtual tip machine" and an object of police analysis to an integral part of developments in institutional strategies of law enforcement. The incorporation of social media logic enables police to more directly influence and determine the timing and delivery of their institutional messages, including messages to members of the public on various authorized police department social media accounts. Increasingly, because of these ongoing developments, journalists and news media organizations are less necessary to filter institutional police messages. The police use of social media for dissemination of institutional messages provides more control to police organizations in some ways. This is not meant to suggest, however, that other traditional media forms have lost their significance. They most certainly have not. Rather, law enforcement agencies, as this book has demonstrated, increasingly tailor their content, messages, and institutional strategies to conform to the logic of social media together with the principles of media logic. These logics operate as mutually reinforcing in relation to policing practices that incorporate media. The television program *COPS*, currently in its twenty-eighth season, illustrates the point.

Police continue to adjust content to fit the format of *COPS* in the same manner as they did in 1989 when the program first aired on television (see Doyle 2003, 32–63). For example, viewers can now stream full episodes of *COPS* online. Official police recruitment advertisements appear alongside episodes on the *COPS* webpage. The *COPS* Facebook page (facebook.com/copstv) and Twitter handle (@COPSTV) also appear embedded in each streamed episode. While "social media platforms can neither take credit nor blame for single-handedly transforming social processes" (Van Dijck and Poell 2013, 11), social media do, however, augment the principles of media logic. Together these processes contribute to developments in the transformation of social order in the form of police social control efforts.

Complementing previous research (Doyle 2003; Ericson, Baranek, and Chan 1989; Ericson and Haggerty 1997; Fishman 1978, 1980, 1981; Manning 2003b), this book has us reflect upon a broader range of considerations. We have seen how police now use social media platforms to advance their own institutional agenda (Meijer and Thaens 2013) on the one hand, and how social media challenges police authority on the other, the conditions of which make police increasingly vulnerable and publicly accountable (Goldsmith 2010). These developments seem to have encouraged police agencies to further incorporate features of media logic into policing practices. Body cameras worn on officers that video record public–police interactions are an extension of police use of media, such as police cruiser–mounted dashboard cameras (Brown 2013). Contemporary policing efforts on social media should not underscore the importance of the influence of mass media in police practice or police control tactics (Doyle 2003; Ericson, Baranek, and Chan 1989; Sacco 1995). Rather, as this book demonstrates, contemporary developments on social media not only amplify existent police practices, but also greatly expand these practices to include many other police activities beyond the parameters of "crime work" (Ericson 1982). Future research could expand our understanding of how actual practices challenge, transform, and reinforce police work.

POLICE ACTIVITY ON SOCIAL MEDIA

Police activity on social media influences and shapes what subject matter will be discussed and not discussed. Police can actively and directly choose when and how discussions will occur, from select preplanned thematic discussions (e.g., #copchat), or those general discussions that might follow press releases on social media. Police efforts to control public discussion vary considerably, from dialoguing with citizens in online chats, to posting sports-themed tweets, and soliciting crime information. All of this information is generated from activity on social media. Police engagement with online users helps channel the direction of user conversations to better meet the interests of police.

The purpose of this book has been to help illustrate the process by which police activity conforms to the logic of social media, and how this activity contributes to recognizable institutional changes in law enforcement. Initially, institutional changes begin with police attraction to social media. Early police attention to social media in North America was largely for investigative purposes, and specifically looked into user activities on MySpace based on the belief of the susceptibility of youth users to the perceived prevalence of online sexual predators (boyd 2014; Marwick 2008). A shift occurs when crimes that unfold online draw the attention of an audience that is collectively and interactively able to contribute to the meaning-making process (Schneider and Trottier 2012). This activity provides definitions of situations (Waller 1970). These definitions operate alongside and, in some cases, can compete with police definitions.

Beyond police–citizen engagement and the delivery of institutional messages on social media, the tens of thousands of police posts on social media platforms represent new forms of police "organization products" (Manning 2003b). These products become useful to journalists as an even more widespread and encompassing feature of contemporary media culture. Police posts on social media, both authorized and private, can become *the* story in news media. These products are the next phase in the postjournalism era, where "issues that journalists report about are themselves products of media" (Altheide and Snow 1991, x). Often stories develop around unofficial officer posts that reflect unfavorably on police. These kinds of stories and police posts illustrate the continued necessity for police agencies to develop social media guidelines that outline expectations of police practices. The development of social media guidelines can occur years after establishing an official presence on social media. The points to stress are that *content and form have become equally important,* and media logic is the guiding principle. The following statement by Police Lieutenant Glen Mills of the Burlington, Massachusetts, police department is an example. Mills noted that a core benefit of social media for the police is to "build an audience year-round" (Qtd. in Laidler 2014).

Building an interactive audience has become another important feature of contemporary police activity on social media. The objective is to attract large followings where audience members on social media might reside far outside police jurisdictional boundaries. The goal becomes the cultivation of an interactive public audience, one that police may not directly serve. Such endeavors contribute to the more widespread reproduction of police legitimacy, the consequences of which are not yet fully known or understood. Police messages in the form of interaction with online users are provided in specific request to audience members within

and beyond official jurisdiction (e.g., #copchat) where a social media audience "market share" emerges among law enforcement agencies. Re-
markably, police agencies now compete for the largest market share on
social media. A list of the top police agencies in the United States (exclud-
ing federal agencies) with the most followers on Twitter is generated
quarterly by the IACP Center for Social Media. One of the leaders is the
Massachusetts State Police, which boasts more than 170,000 Twitter followers.

THE E-AUDIENCE AND THE DEFINITION OF THE SITUATION

Authorized police posts increasingly serve to attract and develop an e-
audience, on the one hand, whereas unofficial police posts, on the other
hand, may serve as a new form of entertainment value. Headlines in
news reports such as "Officer's Facebook Post Sparks Uproar" (*USA Today*, July 23, 2012) help illustrate the point. Moreover, this development
continues to highlight the role of news media itself as a basic part of the
audience. The concept of the audience as a feature of media logic is not a
new one. I have shown, however, that the format and accompanying
logic of social media now allows audience members to take both *passive*
and *active* roles. This important distinction explored in the preceding case
studies contributes to the scholarship on the transformative dimension of
media and the definition of the situation by detailing the importance of
this process in direct relation to the audience response.

Because social media have complicated the control of the definition of
the situation for police, the incorporation of social media logic as an
institutional policing strategy develops as a necessity to more effectively
manage social situations as conducive to the police mandate: "The
contradiction and expansion of an occupation's *mandate* reflects the concerns society has with the services it provides, with its organization, and
with its effectiveness" (Manning 1978, 191, emphasis original). As noted
in *Social Media: A Valuable Tool with Risks*, a joint publication of the Major
Cities Chiefs Associates, Major Counties Sheriffs Associates, and the Federal Bureau of Investigation National Executive Institute Associates, "The
public increasingly expects that the police and other emergency service
providers will be using SM [social media] to respond to calls for service"
(2013, 7). The three preceding case studies empirically demonstrate this
public expectation. Moreover, as previous scholarship has shown, "One's
competence is often judged by communicative performance, but this performance increasingly involves the direct or indirect manipulation of information technology and communication formats" (Altheide 1995, 7).
These case studies also contribute to our understandings of how our
social activities are mediated through information technology (Couch
1984; Carey 1989; Meyrowitz 1985; Altheide and Snow 1991; Altheide

1995). Social media logic applies to the institutional activities and the perceived competence of law enforcement agencies as a basic feature of police image work (Ericson 1982; Mawby 2002b; Schneider 2015b).

INSTITUTIONALIZATION OF SOCIAL MEDIA LOGIC

As social media platforms continue to emerge, develop, and flourish, the use and popularity of some platforms will wax (e.g., Facebook) and wane (e.g., MySpace); however, the logic will remain. The expansion of this logic has altered institutional police practices such that police agencies conform and tailor content to the principles of social media. The significance of media logic on police practice is not new, as detailed by Doyle (2003) and others; rather, my approach has been to identify some of these changes in public police practices that have incorporated social media. As the chapters reveal, police practices have been transformed through media logic. An example includes the creation of news media press releases as a standard police practice and the employment of "press officers who generate contacts with the media and 'feed them'" (Ericson 1982, 8; see also Mawby 2002b). A development of this police practice now includes the employment of social media officers. These officers continue to generate contacts with media, but these practices expand to also include citizens, other police officers, departments, and so on, on social media platforms.

The institutional practices of issuing carefully crafted press releases to news media remain unaltered. Research has shown that these messages contribute to a "social awareness of crime" (Fishman 1978). This practice continues on social media platforms but in a more widespread manner. For instance, numerous crime posts on each of the more than one hundred authorized TPS social media accounts might be posted in just a single day. Additional posts on authorized TPS accounts that would have never been featured in news media, such as those far beyond the parameters of crime, like self-parodying officers tweeting about donuts, also occur. Official police posts on social media augment institutional police messages in new ways that are expansive both in terms of messages across the media landscape *and* of the police mandate (Manning 1978). These posts now occur in smaller increments, over a much shorter duration of time, making the dissemination of institutional messages much faster. Indeed, because of the sheer quantity of crime-themed posts on social media platforms, awareness of crime becomes much more pervasive as might *culture as crime* "as cultural and criminal practices in contemporary social life" become increasingly spotlighted on social media (Ferrell 1995, 25). The social awareness of the police brand also has a far greater reach across the communication spectrum.

A growing area of concern for police is the speed at which information can travel and call into question police conduct in advance of authorized statements made by police and law enforcement. This is the case with information that might discredit the institution, as chapter 5 explored. A key finding detailed in this book reveals that news media reports continue to remain an important and vital source cited by users online to validate materials on social media (see also Schneider 2015b). News media continue to remain important for the reproduction of police legitimacy (see Mawby 2002b). Nevertheless, it has become apparent that police can no longer afford to sit idly by as user-generated content, especially in relation to crime stores (Chermak 1995), continues to circulate online in absence of police contextualization.

POLICE STRATEGIES ON SOCIAL MEDIA

One basic police strategy that continues to be replicated across law enforcement agencies around the world is to provide dedicated official social media accounts to frontline officers. Evidence suggests that frontline officer posts tend to be of a less serious nature and sometimes irrelevant to police work. Such posts are less likely to be viewed as damaging or problematic to the institutional image and message and, perhaps as importantly, are not necessarily required to be vetted by senior administrators. This accounts for speed of message dissemination on social media on the one hand, and officer discretion on the other hand, since police work "is a highly discretionary activity" (Manning 1997, 146). This police strategy on social media creates less bureaucratic oversight of discretionary activities while fostering the appearance of competence and relevance. Activity on social media also enables police to post and validate crime information (Heverin and Zach 2010).

MEDIATED REALITY AND POLICE AUTHORITY

"Experience is mediated" (Altheide and Snow 1991, 250); this has never before been so readily apparent. To be social is to be mediated. This book helps contribute to media sociology in general and specifically contributes to the scholarship that explores how media formats connect with institutional changes in policing. The cases presented here help reveal how developments in social media logic have an influence over policing and police work. This book details this process as it relates to selected circumstances that contribute to developments in social assumptions and connects these changes to the institutional incorporation of the processes by which messages are constructed within social media formats.

Previous scholarship has focused on oligopolistic media, notably television, and how these media contribute to the development of various "sense making strategies" (Altheide and Snow 1991, 241). Mass media, but television in particular, has been recognized as a powerful and potent medium, as a major social influence to the manner in which social situations are understood and defined (Meyrowitz 1985). Television media may serve as the most important catalyst of the incorporation of media logic within particular social institutions (Altheide and Snow 1991). A largely underexplored development in this area of scholarship includes how social media platforms contribute to institution changes. Social media are an emergent category of media forms that enable the simultaneous production and dissemination of culture. This important development expands the numerous key players of law enforcement in the production of socially constructed reality and social order.

While evidence suggests that the mass media landscape is drastically shifting in response to the influence of social media, including the role of policing and police work in this process (Lee and McGovern 2014), an important revelation made in this book is that dominant media have not lost their stake in relation to institutional police legitimacy. The case of alleged rioter Brock Anton, discussed briefly in chapter 3, perhaps best illustrates the continued importance of traditional news media in relation to institutional police legitimacy. Anton's post was arguably the most infamous one made to social media site Facebook during the 2011 Vancouver riot (Schneider 2015c). His post read, "Maced in the face, hit with Batton, tear gassed twice, 6 broken fingers, blood everywhere, punched a fucken pig in the head with riot gear on knocked him to the ground, through [*sic*] the jersey on a burning cop car flipped some cars, burnt some smart cars, burnt some cop cars, I'm on the news . . . One word . . . History :) :) :)." The post, sometimes in its entirety and sometimes only in excerpts, was included in dozens of national print, radio, and television stories covering the 2011 Vancouver riot. Online users circulated the post and commented on it extensively. Speculation continued for months following the riot about Anton's whereabouts, including the status of his arrest and conviction for his proudly "admitted" crimes. The "legitimate" response offered by police to this widespread speculation was provided more than a year later in a *Globe and Mail* article that absolved Anton of wrongdoing in connection with the above post. The headline of the report read, "Riot 'Lightning Rod' Goes Uncharged; Number of Accused Now 156, Police Say, But Believe Man Who Boasted on Facebook Actually Not As Involved As He Claimed." In reference to Anton's post, Constable Brian Montague, a Vancouver Police Department spokesperson, noted, "[We] have investigated him extensively and if we found him doing the things he said he did, there's no doubt in my mind that we would be requesting charges on him.[1] You kind of have to read between the lines there. *He was obviously down there that night. But he's not doing the things he says he is*" (Dhillon 2012b, emphasis mine). This example, as well as others in this book, demonstrates that the definition of the situation continues to rely largely on official information published and circulated through news media reports, rather than on social media platforms. Such official police statements and validation, however, would have not been necessary had it not been for Anton's offending post on social media in the first place. Definitions of socially constructed reality can now be created online, in real time, and parallel to those in face-to-face situations. This process reveals how social media platforms contribute to important contextual changes in relation to the meaning-making process (Schneider 2015a). It was Anton himself who in fact made the post in a specific context, and, had it been made in another context, its meaning may have been interpreted in an entirely different manner. Because social actions are now observable and subject to continual negotiation and reinterpretation by possibly hundreds of millions of people online, a key concern becomes recognizing how these interpretations and interactions together contribute to the definition of the situation and the role that police as authorities play in this process. Getting a better sense of this social process as it continues to develop in Canada and around the world is part of the future work on the subject.

The case studies developed in each chapter detail some recognizable changes in the organization of policing in Canada that can be attributed to social media. The chapters also reveal shared assumptions that citizens have of the institution of policing, such as the expansion of police work to incorporate the use of social media in a variety of capacities, including crime announcements, traffic updates, and expanded forms of community engagement and social control. The most dominant public assumption is the *expectation* that police *actively* and *efficiently* use social media. These platforms are much more than information sources for police agencies. This book has illustrated that, in addition to news media, social media are increasingly pivotal to police agencies as authorities to make convincing claims to the public.

This book has illustrated the application of the principles of media logic in the digital age in relation to how changes in communication formats have an impact on events, actions, and interactions. Each of the preceding chapters has offered specific qualifications about social interactions, constraints, and formats that, taken together, collectively show some of the ways that these criteria alter contemporary police work. A basic theoretical contribution of the book is the demonstration of the significance of both form *and* content in social media logic. This book investigated the content of social media posts in order to understand the logic of social media, which influences the content of policing and police work online. A methodological contribution of the book is that it provides a framework to define, organize, and examine social media documents and their consequences for organizational practices.

Work in these areas remains to be done. Future research might address how communication technologies and social media logic challenge, transform, and reinforce institutionalized social control strategies across other institutions in social life. Additionally, other qualitative methodologies and comparative research efforts building on existent scholarship (Bayley 1990; Brodeur 2010; Mawby 2014) would help further our understanding of policing and police work on social media. Ethnographies of policing remain necessary to understand new developments in private policing as well as innovations in policing technology (Manning 2014). An ethnographic approach toward policing (Marks 2004) and fieldwork data could surely enhance our understanding of the everyday practices and sense-making processes of officer use of social media, for example, beyond existent work in this area (e.g., Brown 2013, 2015; see also Manning 2015). These data would complement the results of this research by illuminating individual officer reflections or interpretations about social media use. I look forward to such future efforts.

NOTE

1. In British Columbia, police recommend charges and Crown counsel makes the decision whether or not to lay charges based on the likelihood of conviction.

Bibliography

A&M Records, Inc. v. Napster, Inc. 239 F.3d 1004. 2001. http://copyright.gov/fair-use/summaries/a&mrecords-napster-9thcir2001.pdf.

Accenture. 2012. "Why Citizens Demand More Social Media in Law Enforcement." *Mashable*. Accessed August 7, 2014. http://mashable.com/2012/10/09/police-social-media/.

———. 2013. *Preparing Police Services for the Future: Six Steps toward Transformation.* Accenture. https://www.accenture.com/keen/~/media/Accenture/Conversion-Assets/DotCom/Documents/Global/PDF/Industries_7/Accenture-Preparing-Police-Services-Future.pdf.

———. 2014. *How Can Digital Police Solutions Better Serve Citizens' Expectations? Accenture Citizen Pulse Survey on Policing 2014.* Accenture. http://www.accenture.com/SiteCollectionDocuments/PDF/Accenture-How-Can-Digital-Police-Solutions-Better-Serve-Citizens-Expectations.pdf.

Adolf, Marian, and Cornelia Wallner. 2011. "The Wikileaks Affair in German Media: An Analysis of a Discursive Indignation." Istanbul, Turkey: Kadir Has University, International Association for Media and Communication Research.

Agrell, Siri. 2006. "Troubled Kids Gravitating to Vampire Site: Several Violent Crimes in Canada Tied to Network." *National Post*, September 15.

Alamenciak, Tim, and Hamida Ghafour. 2013. "Who Was Sammy Yatim?" *Toronto Star*, August 24. http://www.thestar.com/news/crime/2013/08/24/who_was_sammy_yatim.html.

Allen, Kate. 2011. "Twitter Cop Gets Police Work Done—140 Characters at a Time." *Toronto Star*, August 6. http://www.thestar.com/news/gta/2011/08/06/twitter_cop_gets_police_work_done_140_characters_at_a_time.html.

Altheide, David L. 1976. *Creating Reality: How TV News Distorts Events.* Thousand Oaks, CA: Sage.

———. 1995. *An Ecology of Communication.* Hawthorne, NY: Aldine de Gruyter.

———. 2000. "Identity and the Definition of the Situation in a Mass Mediated Context." *Symbolic Interaction* 32:1–27.

———. 2002. *Creating Fear: News and the Construction of Crisis.* New York: Aldine de Gruyter.

———. 2004. "Ethnographic Content Analysis." In *The SAGE Encyclopedia of Social Science and Research Methods*, edited by M.S. Lewis-Beck, Alan Bryman, and Tim Futing Liao, 325–26. Thousand Oaks, CA: Sage.

———. 2006. *Terrorism and the Politics of Fear.* Lanham, MD: Alta Mira Press.

———. 2014. *Media Edge: Media Logic and Social Reality.* New York: Peter Lang.

Altheide, David L., and John Johnson. 1980. *Bureaucratic Propaganda.* Boston: Allyn & Bacon.

Altheide, David L., and Christopher J. Schneider. 2013. *Qualitative Media Analysis.* 2nd ed. Thousand Oaks, CA: Sage.

Altheide, David L., and Robert P. Snow. 1979. *Media Logic.* Newbury Park, CA: Sage.

———. 1988. "Toward a Theory of Mediation." In *Communication Yearbook 11*, edited by J. A. Anderson, 194–223. Newbury Park, CA: Sage.

———. 1991. *Media Worlds in the Postjournalism Era.* Hawthorne, NY: Aldine de Gruyter.

Arvanitidis, Tania. 2013. "From Revenge to Restoration: Evaluating General Deterrence as a Primary Sentencing Purpose for Rioters in Vancouver, British Columbia."
MA thesis, Simon Fraser University, Burnaby, BC. http://summit.sfu.ca/item/13584.

Asp, Kent. 1986. *Mäktiga massmedier: Studier i politisk opinionsbildning* (Powerful mass
media: Studies in political opinion-formation). Stockholm: Akademilitteratur.

Associated Press. 2005. "Auto Club First 27 in Message Board Crackdown." *USA
Today*, August 6. http://usatoday30.usatoday.com/tech/news/2005-08-06-posters-
fired_x.htm.

Association of Chief Police Officers (ACPO). 2013. *Guidelines on the Safe Use of the
Internet and Social Media by Police Officers and Police Staff.* London: National Police
Chiefs' Council. http://www.btp.police.uk/pdf/FOI%20Response%20319-
14%20ACPO%20Guidance.PDF.

Auer, Doug. 2012. "Finest on Facebook." *New York Post*, June 14. http://nypost.com/
2012/06/14/finest-on-facebook/.

Bahney, Anna. 2006. "Don't Talk to Invisible Strangers." *New York Times*, March 9.

Baker, Stephanie Alice. 2011. "The 'Mediated Crowd': New Social Media Forms and
New Forms of Rioting." *Sociological Research Online* 16 (4). http://
www.socresonline.org.uk/16/4/21.html.

———. 2012a. "From the Criminal Crowd to the 'Mediated Crowd': The Impact of
Social Media on the 2011 English Riots." *Safer Communities* 11 (1): 40–49.

———. 2012b. "Policing the Riots: New Social Media as Recruitment, Resistance, and
Surveillance." In *The English Riot of 2011: A Summer of Discontent*, edited by Daniel
Briggs, 169–90. East Sussex: Waterside Press.

———. 2014. *Social Tragedy: The Power of Myth, Ritual, and Emotion in the New Media
Ecology.* New York: Palgrave Macmillan.

Bakker, J.I. (Hans). 2007. "Definition of the Situation." In *The Blackwell Encyclopedia of
Sociology*, vol. 3, edited by George Ritzer, 991–92. Malden, MA: Oxford Blackwell
Publishing.

Banton, Michael. 1964. *The Policeman in the Community*. London: Tavistock.

Barnett, Emma. 2011. "Facebook Dominance Forces Rival Networks to Go Niche."
Telegraph, January 12. http://www.telegraph.co.uk/technology/MySpace/8253091/
Facebook-dominance-forces-rival-networks-to-go-niche.html.

Bayley, David H. 1990. *Patters of Policing: A Comparative International Analysis.* New
Brunswick, NJ: Rutgers University Press.

———. 1994. *Policing for the Future*. New York: Oxford University Press.

Bazarova, Natalya N. 2012. "Public Intimacy: Disclosure Interpretation and Social
Judgments on Facebook." *Journal of Communication* 62:815–32.

Becket, Katherine, and Theodore Sasson. 2004. *The Politics of Injustice: Crime and Punishment in America.* Thousand Oaks, CA: Sage.

Berger, Peter, and Thomas Luckman. 1967. *The Social Construction of Reality: A Treatise
in the Sociology of Knowledge.* Harmondsworth, UK: Penguin Books.

Bergeson, Albert, and Max Herman. 1998. "Immigration, Race and Riot: The 1992 Los
Angeles Uprising." *American Sociological Review* 63 (1): 39–54.

Bertot, John C., Paul T. Jaeger, and Justin M. Grimes. 2010. "Using ICTs to Create a
Culture of Transparency: E-government and Social Media as Openness and Anti-
Corruption Tools for Societies." *Government Information Quarterly* 27 (3): 264–71.

Bittner, Egon. 1970. *The Functions of Police in Modern Society*. Washington, DC: U.S.
Government Printing Office.

Black, Donald. 1983. "Crime as Social Control." *American Sociological Review* 48 (1):
34–45.

Blumer, Herbert. 1939. "Collective Behavior." In *An Outline of Principles of Sociology*,
edited by R.E. Park, 219–80. New York: Barnes & Noble.

———. 1951. "Collective Behavior." In *New Outline of the Principles of Sociology*, edited
by A.M. Lee, 166–222. New York: Barnes & Noble.

———. 1969. *Symbolic Interaction: Perspective and Method*. Englewood Cliffs, NJ: Prentice-Hall.

Boesveld, Sarah. 2013. "Blair Vows to Probe Teen Killing; 'Serious Concerns'; Police Fired Multiple Shots at Armed Victim." *National Post* (Toronto), July 30.

boyd, danah. 2014. *It's Complicated: The Social Lives of Networked Teens.* New Haven, CT: Yale University Press.

boyd, danah m., and Nicole B. Ellison. 2007. "Social Network Sites: Definition, History, and Scholarship." *Journal of Computer-Mediated Communication* 13 (1): 210–30.

boyd, danah, and Eszter Hargittai. 2010. "Facebook Privacy Settings: Who Cares?" *First Monday* 15 (8). http://firstmonday.org/article/viewArticle/3086/2589.

boyd, danah, Scott Golder, and Gilad Lotan. 2010. "Tweet, Tweet, Retweet: Conversational Aspects of Retweeting on Twitter." HICSS-43. IEEE: Kauai, HI, January 6. http://www.danah.org/papers/TweetTweetRetweet.pdf.

Braidwood, Thomas R. 2010. *Why? The Robert Dziekanski Tragedy.* Victoria, BC: Braidwood Commission on the Death of Robert Dziekanski.

Bright Planet. 2013. "How Police Departments Use Twitter." (Infographic). *Connected-COPS*, April 23. http://connectedcops.net/infographic-how-police-departments-use-twitter/.

Briscoe, Daren. 2006. "'Netbangers,' Beware." *Newsweek*, March 13.

Brissett, Dennis, and Charles Edgley. 1990. *Life as Theater: A Dramaturgical Source Book.* New York: Aldine de Gruyter.

British Columbia Police Commission. 1994. *Report on the Riot That Occurred in Vancouver on June 14–15, 1994.* Victoria, BC: British Columbia Police Commission. Accessed February 14, 2015. http://beta.images.theglobeandmail.com/archive/01289/Report_on_the_1994_1289365a.pdf.

Brodeur, Jean-Paul. 2010. *The Policing Web.* New York: Oxford University Press.

Brown, Gregory Roy. 2013. "The Blue Line on Thin Ice: Police Use of Force in the Era of Cameraphones, 'Citizen Journalism,' and YouTube." MA thesis, Carleton University, Ottawa, ON. https://curve.carleton.ca/system/files/theses/27515.pdf.

———. 2015. "The Blue Line on Thin Ice: Police Use of Force Modifications in the Era of Cameraphones and YouTube." *British Journal of Criminology.* doi:10.1093/bjc/azv052.

Brunty, Joshua, and Katherine Helenek. 2015. *Social Media Investigations for Law Enforcement.* New York: Routledge.

Bui, Lynh. 2014. "Prince George's Police Plan to Live-tweet Prostitution Sting." *Washington Post*, May 3. http://www.washingtonpost.com/local/crime/prince-georges-police-plan-to-live-tweet-prostitution-sting/2014/05/03/d88719c8-d22e-11e3-937f-d3026234b51c_story.html.

Burke, Kenneth. 1965. *Permanence and Change.* Chicago: Bobbs-Merrill.

Burrows, Tim. 2012a. "About Me." http://about.me/timburrows.

———. 2012b. *Walking the Social Media Beat: Twitter Guide for Police and Law Enforcement.* Self-published e-book. http://walkingthesocialmediabeat.com/2012/10/09/the-twitter-book-for-police-law-enforcement/.

———. 2014. "The Toronto Police: Building a More Social Service." *Hootsuite*, July 30. http://blog.hootsuite.com/toronto-police-building-social-service/.

Cain, Maureen. 1973. *Society and the Policeman's Role.* London: Routledge.

Campeau, Holly. 2015. "'Police Culture' at Work: Making Sense of Police Oversight." *British Journal of Criminology.* doi:10.1093/bjc/azu093.

Canadian Press. 1994. "News Briefing: CBC Withholds Raw Tape of Riot." *Globe and Mail*, June 23.

———. 2013. "Alberta Deaths Put Police's Use of Lethal Force under Renewed Scrutiny." *Globe and Mail*, August 4. http://www.theglobeandmail.com/news/national/alberta-rcmp-fire-guns-during-roadside-stop-for-second-time-in-under-a-week/article13589023/.

Canadian Broadcasting Corporation (CBC). 2011a. "Mounties Miffed about Fake Twitter Account." *CBC News*, November 17. http://www.cbc.ca/news/canada/saskatchewan/mounties-miffed-about-fake-twitter-account-1.1123401.

———. 2011b. "Public Shaming." *The Current*, June 21. http://www.cbc.ca/thecurrent/episode/2011/06/21/public-shaming/.

———. 2011c. "A Tale of Two Riots." *CBC News*, June 16. http://www.cbc.ca/news/canada/a-tale-of-two-riots-1.1079520.

———. 2011d. "Vancouver Police Arrest More Than 100 in Riot." *CBC News*, June 16. http://www.cbc.ca/news/canada/british-columbia/story/2011/06/16/bc-riot-thursday.html.

Canadian Wireless Telecommunications Association (CWTA). 2015. "Wireless Phone Subscribers in Canada." (Table). Accessed December 10. http://www.cwta.ca/wp-content/uploads/2015/11/SubscribersStats_en_2015_Q3.pdf.

Carey, James. 1989. *Communication as Culture: Essays on the Media and Society*. Boston: Unwin Hyman.

Carey, James, ed. 1987. *Media, Myths and Narratives*. Newbury Park, CA: Sage.

CBC TV (Vancouver). 2011. *Vancouver News at 11*. Television broadcast, June 15.

Cernetig, Miro. 1994. "Why Media Didn't Volunteer Their Riot Photos." *Globe and Mail*, July 15.

Clark, Christy. 2011. "An Open Letter to British Columbians from Premier Clark." Statement from the Office of the Premier, British Columbia, June 16. http://www2.news.gov.bc.ca/news_releases_2009-2013/2011PREM0070-000730.htm.

Chan, Janet B. L. 2001. "The Technology Game: How Information Technology is Transforming Police Practice." *Criminology and Criminal Justice* 1 (2): 139–59.

Chang, Lennon Y. C. 2013. "Formal and Informal Modalities for Policing Cybercrime across the Taiwan Strait." *Policing and Society* 23 (4): 540–55.

Chermak, Steven. 1995. "Image Control: How Police Affect the Presentation of Crime News." *American Journal of Police* 14 (2): 21–43.

Chermak, Steven, and Alexander Weiss. 2005. "Maintaining Legitimacy Using External Communication Strategies: An Analysis of Police-Media Relations." *Journal of Criminal Justice* 33:501–12.

Chung, Jamie. 2012. "Coquitlam Mounties Aim to be Project X Party-Poopers." BC RCMP. Accessed January 15, 2014. http://bc.rcmpgrc.gc.ca/ViewPage.action?siteNodeId=319&languageId=1&contentId=26769.

City of Vancouver. 2011. City of Vancouver Internal Review of the 2011 Stanley Cup Riot. Vancouver, BC: City of Vancouver. http://council.vancouver.ca/20110906/documents/specAppendixA1.pdf.

Cohen, Heidi. 2011. "30 Social Media Definitions." *Heidi Cohen: Actionable Marketing Guide* (blog), May 9. http://heidicohen.com/social-media-definition/.

Cohen, Stanley. 1985. *Visions of Social Control: Crime, Punishment and Classification*. Cambridge: Polity Press.

Community Oriented Policing Services (COPS) and Police Executive Research Forum. 2013. *Social Media and Tactical Considerations for Law Enforcement*. Washington, DC: Office of Community Oriented Policing Services, U.S. Department of Justice.http://policeforum.org/library/technology/SocialMediaandTacticalConsiderationsforLawEnforcement.pdf.

ConnectedCOPS. 2011. "The Toronto Police Service Launches Social Media Program." *ConnectedCOPS*, July 25.http://connectedcops.net/2011/07/25/the-toronto-police-service-launches-social-media-program/.

Cooley, Charles H. 1983. "Looking Glass Self." In *Human Nature and the Social Order*, edited by C. H. Cooley, 183–85. New Brunswick, NJ: Transaction Publishers.

Couch, Carl J. 1984. *Constructing Civilizations*. Greenwich, CT: JAI.

Couch, Carl J., David Maines, and Shing-Ling Cheng. 1996. *Information Technologies and Social Orders*. New York: Aldine de Gruyter.

Council of Canadian Academies. 2014. *Policing Canada in the 21st Century: New Policing for New Challenges*. Ottawa, ON: The Expert Panel on the Future of Canadian Policing Models, Council of Canadian Academies.

Critcher, Chas. 2003. *Moral Panics and the Media*. Philadelphia, PA: Open University Press.

Crump, Jeremy. 2011. "What Are the Police Doing on Twitter? Social Media, the Police, and the Public." *Policy and Internet* 3 (4): 1–27.

Dahlgren, Peter. 1996. "Media Logic in Cyberspace: Repositioning Journalism and Its Publics." *The Public* 3:59–72.

Daily News (Nanaimo, BC). 2010. "Nanaimo Mountie Red-Faced over Public Facebook Posts." March 27. http://www.canada.com/story.html?id=14b1672a-6ae7-44d8-acda-fbe6d5839f1c.

Dame, Jonathan. 2014. "Will Employers Still Ask for Facebook Passwords in 2014?" *USA Today*, January 10.

Dhillon, Sunny. 2011. "When Rioters Trashed Vancouver, Twitter Fanned the Flames and Gathered the Evidence." *Globe and Mail*, December 21. http://www.theglobeandmail.com/news/british-columbia/when-rioters-trashed-vancouver-twitter-fanned-the-flames---and-gathered-the-evidence/article4182089/.

———. 2012a. "Police Arrest No. 1 Stanley Cup Riot." *Globe and Mail*, September 27. http://www.theglobeandmail.com/news/british-columbia/police-arrest-no1-target-in-stanley-cup-riot/article4570380/.

———. 2012b. "Stanley Cup 'Lightning Rod' Goes Uncharged." *Globe and Mail*, September 25. http://www.theglobeandmail.com/news/british-columbia/stanley-cup-riot-lightning-rod-goes-uncharged/article4568504/.

Dowler, Kenneth. 2003. "Media Consumption and Public Attitudes Toward Crime and Justice: The Relationship between Fear of Crime, Punitive Attitudes, and Perceived Police Effectiveness." *Journal of Criminal Justice and Popular Culture* 10 (2): 109–26.

Doyle, Aaron. 2003. *Arresting Images: Crime and Policing in front of the Television Camera.* Toronto: University of Toronto Press.

———. 2006. "How Not To Think about Crime in the Media." *Canadian Journal of Criminal Justice* 48 (6): 867–85.

Drake, Laura, 2009. "A Badge, a Gun and a Facebook Profile; Police Learn to Tap Social-Networking Sites." *Edmonton Journal*, March 29.

Durkheim, Emile. 1966. *The Division of Labour in Society.* Translated by G. Simpson. Toronto: Collier-Macmillan.

———. (1895) 1982. *The Rules of Sociological Method.* New York: Free Press.

Elash, Anita. 2011. "Summer Time and the Living Ain't Easy; Riots, Recession Rears and Even Rampaging Bugs Marked—and Marred—the Season for Many." *Globe and Mail*, September 3.

Ericson, Richard V. 1982. *Reproducing Order: A Study of Police Patrol Work.* Toronto: University of Toronto Press.

———. 1991. "Mass Media, Crime, Law, and Justice: An Institutional Approach." *British Journal of Criminology* (31): 219–49.

———. 1995. "The News Media and Accountability in Criminal Justice." In *Accountability for Criminal Justice*, edited by Phillip C. Stenning, 135–61. Toronto: University of Toronto Press.

Ericson, Richard V., and Kevin Haggerty. 1997. *Policing the Risk Society.* Toronto: University of Toronto Press.

Ericson, Richard V., Patricia M. Baranek, and Janet B.L. Chan. 1987. *Visualizing Deviance: A Study of News Organization.* Toronto: University of Toronto Press.

———. 1989. *Negotiating Control: A Study of News Sources.* Toronto: University of Toronto Press.

———. 1991. *Representing Order: Crime, Law, and Justice in the News Media.* Toronto: University of Toronto; Milton Keynes: Open University Press.

Federal Bureau of Investigation (FBI). 2006. "Social Networking Sites: Online Friendships Can Mean Offline Peril." *FBI Stories*, April 3. http://www.fbi.gov/news/stories/2006/april/social_networking040306.

Fenton, Mark. 2013. "In the Social Media Age, Can Justice Be Served?" *UBC Dialogues: Vancouver*, March 13. Podcast audio. https://www.alumni.ubc.ca/2013/events/dialogues/ubc-dialogues-vancouver-social-media-justice/.

Ferrell, Jeff. 1995. "Culture, Crime, and Cultural Criminology." *Journal of Criminal Justice and Popular Culture* 3 (2): 25–42.

Ferrell, Jeff, Keith Hayward, Wayne Morrison, and Mike Presdee, eds. 2004. *Cultural Criminology Unleashed*. London: GlassHouse; Portland, OR: Routledge-Cavendish.

Finnemann, Niels Ole. 2011. "Mediatization Theory and Digital Media." *Communications* 36 (1): 67–89.

Fishman, Mark. 1978. "Crime Waves as Ideology." *Social Problems* 25:531–43.

———. 1980. *Manufacturing the News*. Austin: University of Texas Press.

———. 1981. "Police News: Constructing an Image of Crime." *Urban Life* 9 (4): 371–94.

Fishman, Mark, and Gray Cavender, eds. 1998. *Entertaining Crime: Television Reality Programs*. New York: Aldine de Gruyter.

Fitzpatrick, Meagan. 2013. "10 Voices on the 'Not Criminally Responsible' Reform." *CBC News*, June 6. http://www.cbc.ca/news/politics/story/2013/06/05/polncr-committee-voices.html.

Forrest, Emma. 2005. "Teenagers and the Net." *The Guardian*, December 11. http://www.theguardian.com/technology/2005/dec/11/news.observerfocus.

Fox, Chris. 2014. "Belleville Police Service Investigating 'Inappropriate' Tweets." *CP24 News*, December 9. http://www.cp24.com/news/belleville-police-service-investigating-inappropriate-tweets-1.2139788.

Frey, James H., and Stanley D. Eitzen. 1991. "Sports and Society." *Annual Review of Sociology* 17:503–22.

Fries, Amanda. 2012. "Police Use Social Media to Keep Public Safe, Informed." *Utica Observer-Dispatch* (Mowhak Valley, NY), February 24. http://www.uticaod.com/x565044829/Police-use-social-media-to-keep-public-safe-informed.

Furedi, Frank. 1997. *Culture of Fear: Risk-Taking and the Morality of Low Expectation*. London: Cassell.

Furlong, John, and Douglas J. Keefe. 2011. *The Night the City Became a Stadium: Independent Review of the 2011 Vancouver Stanley Cup Playoff*. Vancouver: Government of British Columbia. http://www2.gov.bc.ca/assets/gov/law-crime-and-justice/about-bc-justice-system/inquiries/report.pdf.

Garrett, Ronnie. 2006. "Catch a Creep: Come on Over to MySpace and You'll Solve Crimes." *Law Enforcement Technology* 33 (11): 10–19.

Gibbs, Jack P. 1977. "Social Control, Deterrence, and Perspectives on Social Order." *Social Forces* 56 (2): 408–23.

Glaberson, William. 2011. "On Facebook, N.Y.C. Police Officers Maligned West Indian Paradegoers." *New York Times*, December 5. http://www.nytimes.com/2011/12/06/nyregion/on-facebook-nypd-officers-malign-west-indian-paradegoers.html?pagewanted=all#commentsContainer.

Glaser, Barney G., and Anselm L. Strauss. 1967. *Discovery of Grounded Theory: Strategies for Qualitative Research*. Chicago, IL: Aldine de Gruyter.

Glassner, Barry. 1999. *The Culture of Fear: Why Americans Are Afraid of the Wrong Things*. New York: Basic Books.

Glauser, Wendy. 2013. "Police Call for More Mental Health Services." *Canadian Medical Association Journal* 185 (17): 1485.

Goffman, Erving. 1959. *The Presentation of Self in Everyday Life*. New York: Doubleday Press.

———. 1974. *Frame Analysis: An Essay on the Organization of Experience*. New York: Harper & Row.

Goldsmith, Andrew. 2010. "Policing's New Visibility." *British Journal of Criminology* 50:914–34.

———. 2015. "Disgracebook Policing: Social Media and the Rise of Police Indiscretion." *Policing and Society* 25 (3): 249–67. doi:10.1080/10439463.2013.864653.

Goldstein, Joseph. 1960. "Police Discretion Not to Invoke the Criminal Process: Low-Visibility Decisions in the Administration of Justice." *Yale Law Journal* 69 (4): 543–94.

Goode, Eriche, and Nachman Ben-Yehuda. 1994. *Moral Panics: The Social Construction of Deviance*. Cambridge, MA: Blackwell.

Goodman, David J., and Wendy Ruderman. 2013. "Police Dept. Sets Rules for Officers' Use of Social Media." *New York Times*, March 28. http://www.nytimes.com/2013/03/29/nyregion/new-york-police-dept-issues-guidelines-for-social-media.html?_r=0.

Gordon, Leonard. 1983. "Aftermath of a Race Riot: The Emergent Norm Process among Black and White Community Leaders." *Sociological Perspectives* 26 (2): 115–35.

Graber, Doris. 1980. *Crime News and the Public*. New York: Praeger.

Greer, Chris, and Eugene McLaughlin. 2010. "We Predict a Riot? Public Order Policing, New Media Environments and the Rise of the Citizen Journalist." *British Journal of Criminology* 50:1041–59.

Grogan, Billy. 2014. "Social Media Created Citizen Investigators." *The Social Media Beat* (blog), *IACP Center for Social Media*, October 24. http://blog.iacpsocialmedia.org/Home/tabid/142/entryid/406/Default.aspx.

Grossberg, Lawrence, Ellen Wartella, D. Charles Whitney, and J. Macgregor Wise. 2006. *Media Making: Mass Media in Popular Culture*. Thousand Oaks, CA: Sage.

Grossman, Lev. 2006. "You—Yes, You—Are TIME's Person of the Year." *Time Magazine*, December 25. http://content.time.com/time/magazine/article/0,9171,1570810,00.html.

Gutierrez, Melinda. 2011. "New Social Media Duties for Dallas Police Officer." *Connected Cops*, April 11. http://connectedcops.net/2011/04/11/new-social-media-duties-for-dallas-police-officer/.

Hartley, Matt. 2008. "Before We Built the Prototype for Twitter, People Didn't Seem That Interested In It. It Was a Real Shock When People Gravitated toward It Very Quickly." *Globe and Mail*, July 23.

Hasham, Alyshah. 2013. "Speedy Bail Hearing Raises Legal Eyebrows." *Toronto Star*, August 21.

Hepp, Andreas. 2011. "Mediatization, Media Technologies and the 'Moulding Forces' of the Media." Paper presented at the International Communication Association Conference, Boston. Accessed August 10, 2014. http://www.andreas-hepp.name/Blog/Eintrage/2011/5/26_Paper_auf_der_ICA-Tagung_und_mehr_files/Hepp.pdf.

———. 2013a. "The Communicative Figurations of Mediatized Worlds: Mediatization Research in Times of the 'Mediation of Everything.'" *European Journal of Communication* 28 (6): 615–29.

———. 2013b. *Cultures of Mediatization*. London: Polity Press.

Hepp, Andreas, and Friedrich Krotz, eds. 2014. *Mediatized Worlds: Culture and Society in a Media Age*. Basingstoke, UK: Palgrave Macmillan.

Hernes, Gudmund. 1978. *For-handlingsekonomi og blandningsadministrasjon*. Edited by G. Hernes. Bergen: Universitetsforlaget.

Heverin, Thomas, and Lisl Zach. 2010. "Twitter for City Police Department Information Sharing." Proceedings of ASIST 2010, October 22–27, Pittsburgh, PA. http://www.asis.org/asist2010/proceedings/proceedings/ASIST_AM10/submissions/277_Final_Submission.pdf.

Hjarvard, Stig Prof. 2013. *The Mediatization of Culture and Society*. New York: Routledge.

Honeycutt, Courtenay, and Susan C. Herring. 2009. "Beyond Microblogging: Conversation and Collaboration in Twitter." Proceedings of the 42nd Hawaii International Conference on System Sciences. http://www.computer.org/csdl/proceedings/hicss/2009/3450/00/03-05-05.pdf.

Howard, Ross. 1995. "No Mistakes Made in Riot: Vancouver Police Review Says Chief Faults 'Technical Problem' in Stanley Cup Fracas." *Globe and Mail*, February 4.

Hrin, Eric. 2011. "Troy Police Facebook Page Use Temporarily Discontinued." *Daily Review* (Troy, PA), November 18.

Hui, Ann. 2013. "Toronto Police Mull Body-Worn Video Cameras for Officers." *Globe and Mail*, October 18. http://www.theglobeandmail.com/news/toronto/body-worn-video-cameras-being-considered-for-toronto-police/article14923131/.

Hunter, Albert. 1985. "Private, Parochial and Public Social Orders: The Problem of Crime and Incivility in Urban Communities." In *The Challenge of Social Control*, edited by Gerald D. Suttles and Mayer N. Zald, 230–42. Norwod, NJ: Ablex Publishers.

Innis, Harold. 1951. *The Bias of Communication*. Toronto: University of Toronto Press.

International Association of the Chiefs of Police. 2013. "Using Social Media as an Investigative Tool." *The Social Media Beat, IACP Center for Social Media Blog*, October 29. http://blog.iacpsocialmedia.org/Home/tabid/142/entryid/315/Default.aspx.

Janowitz, Morris. 1978. *The Last Half-Century: Societal Change and Politics in America.* Chicago: University of Chicago Press.

———. 1991. *On Social Organization and Social Control.* Edited with an Introduction by James Burk. Chicago: University of Chicago Press.

Johns, Mark D., Shing-Ling Sarina Chen, and Laura Terlip, eds. 2014. "Symbolic Interaction and New Social Media." *Studies in Symbolic Interaction* 43.

Jones, Helena. 2007. "Media: Publishing 2.0: The Changing Face of Publishing." *The Guardian*, February 12.

Jones, Trevor, and Tim Newburn. 2002. "The Transformation of Policing? Understanding Current Trends in Policing Systems." *British Journal of Criminology* 42:129–46.

Kappeler, Victor E., and Gary W. Potter. 2005. *The Mythology of Crime and Criminal Justice.* 4th ed. Long Grove, IL: Waveland Press.

Kari, Shannon. 2015. "Forcillo Testifies He Was 'Trained to Win' at Trial for Yatim Shooting Death." *Globe and Mail*, November 25. http://www.theglobeandmail.com/news/toronto/james-forcillo-cross-examined-in-trial-for-killing-of-sammy-yatim/article27509017/.

Kasinsky, Renee Goldsmith. 1995. "Patrolling the Facts: Media, Cops, and Crime." In *Media, Process, and the Social Construction of Crime: Studies in Newsmaking*, edited by Gregg Barak, 203–36. New York: Garland Publishing.

Kauri, Vidya. 2013. "Hundreds Attend 'Justice for Sammy' Protest Against Fatal Streetcar Shooting." *Globe and Mail*, July 29. http://www.theglobeandmail.com/news/toronto/hundreds-of-protesters-march-against-fatal-streetcar-shooting/article13492101/.

Kemp, Joe. 2012. "NYPD Disciplines 17 Cops Who Posted Racist Facebook Comments about West Indian Day Parade." *New York Daily News*, August 23. http://www.nydailynews.com/new-york/nypd-disciplines-17-cops-postedracist-facebook-comments-west-indian-day-parade-article-1.1142642.

Kern, Vera. 2013. "Hunting Criminals on Facebook." *Deutsche Welle*, April 12. http://www.dw.de/hunting-criminals-on-facebook/a-17270778.

Killian, Lewis M. 1984. "Organization, Rationality, and Spontaneity in the Civil Rights Movement." *American Sociological Review* 49:770–83.

Kirkpatrick, David. 2010. *The Facebook Effect: The Inside Story of the Company That Is Connecting the World.* New York: Simon & Schuster.

Kittler, J. 2011. "MySpace." In *Encyclopedia of Social Networks*, edited by G. Barnett, 564–67. Thousand Oaks, CA: Sage.

Klein, Jeff Z., and Ian Austen. 2011. "Hockey Hangover Turns into Riot Embarrassment." *New York Times*, June 16. http://www.nytimes.com/2011/06/17/world/americas/17vancouver.html?_r=2&.

Klockars, Carl B. 1985. *The Idea of Police.* Beverly Hills, CA: Sage.

———. 1988. "The Rhetoric of Community Policing." In *Community Policing: Rhetoric or Reality*, edited by J. R. Green and S. D. Mastrofski, 239–58. New York: Praeger.

Knibbs, Katie. 2013. "How Police Use Social Networks For Investigations. Digital Trends." *Digital Trends*, July 13. http://www.digitaltrends.com/social-media/the-new-inside-source-for-police-forces-social-networks/#ixzz2ooin29Uv.

Koper, Christopher S., Cynthia Lum, and James J. Willis. 2014. "Optimizing the Use of Technology in Policing: Results and Implications from a Multi-Site Study of the Social, Organizational, Behavioural Aspects of Implementing Police Technologies." *Policing: A Journal of Policy and Practice* 8 (2): 212–21.

Kornblum, Janet. 2005. "Teens Wear Their Hearts on Their Blog." *USA Today*, October 31.

Kwak, Haewoon, Changhyun Lee, Hosung Park, and Sue Moon. 2010. "What is Twitter, a Social Network or a News Media?" Proceedings of the 42nd Hawaii International Conference on System Sciences—2009. IEEE: Kauai, HI. http://www. computer.org/csdl/proceedings/hicss/2009/3450/00/03-05-05.pdf.

Laidler, John. 2014. "Police Attracting Crowd on Twitter." *Boston Globe*, February 16. http://www.bostonglobe.com/metro/regionals/north/2014/02/16/billerica-burlington-and-tyngsborough-police-attracting-crowd-twitter/hFgvZv0YiTYk0WJyp28nMP/story.html.

Lanier, Mark M., and Stuart Henry. 1998. *Essential Criminology*. Boulder, CO: Westview Press.

Lasswell, Harold. 1948. "The Structure and Function of Communication in Society." In *The Communication of Ideas: A Series of Addresses*, edited by Lyman Bryson, 37–51. New York: Institute for Religious and Social Studies.

Lavoie, Jennifer A. A., Judy Eaton, Carrie B. Sanders, and Matthew Smith. 2014. "'The Wall is the City': A Narrative Analysis of Vancouver's Post-Riot Apology Wall." In *Symbolic Interaction and New Social Media*, edited by Mark D. Johns, Shing-Ling S. Chen, and Laura A. Terlip, 203–22. Bingley, UK: Emerald Group Publishing.

Lawrence, Regina G. 2000. *The Politics of Force: Media and the Construction of Police Brutality*. Berkeley: University of California Press.

Lee, Robert Mason. 1994. "Vancouver Turns Blind Eye to the Police Role on the Night of Shame." *Globe and Mail*, November 12.

Lee, Murray, and Alyce McGovern. 2013. "Force to Sell: Policing the Image and Manufacturing in Public Confidence." *Policing and Society* 23 (2): 103–24.

———. 2014. *Policing and Media: Public Relations, Simulations, and Communication*. New York: Routledge.

Leishman, Frank, and Paul Mason. 2003. *Policing and the Media: Facts, Fictions, and Factions*. Cullompton, UK: Willan.

Lemert, Edwin M. 1951. *Social Pathology: Systematic Approaches to the Study of Sociopathic Behavior*. New York: McGraw-Hill.

Leukfeldt, Rutger, Sander Veenstra, and Wouter Stol. 2013. "High Volume Cyber Crime and the Organization of the Police: The Results of Two Empirical Studies in the Netherlands." *International Journal of Cyber Criminology* 7 (1). http://www.cybercrimejournal.com/Leukfeldtetal2013janijcc.pdf.

Lewis, Paul, Tim Newburn, Matthew Taylor, Catriona Mcgillivray, Aster Greenhill, Harold Frayman, and Rob Proctor. 2011. *Reading the Riots: Investigating England's Summer of Disorder*. London: The Guardian and the London School of Economics. Accessed November 20, 2014. http://eprints.lse.ac.uk/46297/1/Reading%20the%20riots%28published%29.pdf.

Lieberman, Joel D., Deborah Koetzle, and Mari Sakiyama. 2013. "Police Departments' Use of Facebook Patterns and Policy Issues." *Police Quarterly* 16 (4): 438–62.

Lieberson, Stanley, and Arnold R. Silverman. 1965. "The Precipitants and Underlying Conditions of Race Riots." *American Sociological Review* 30:887–98.

Lindsay, Eathan. 2014. "Social Media as an Outlet for Community Response and Dialogue Following the 2011 Vancouver Stanley Cup Riot." MA thesis, University of Ottawa, Ottawa, ON. http://www.ruor.uottawa.ca/bitstream/10393/31546/1/Lindsay_Eathan_2014_thesis.pdf.

Locke, Laura. 2007. "The Future of Facebook." *Time*, July 17. http://content.time.com/time/business/article/0,8599,1644040,00.html#ixzz2peY0qQDz.

Longley, Anne. 2011. "How Social Media Shaped a Riot Investigation and the Community's Response to the Crimes." Presented to the IACP Annual Conference October 24. Accessed June 23, 2014. http://www.iacpsocialmedia.org/Portals/1/documents/2011%20Conference%20Flyer.pdf.

———. 2013. "#MyDash: Constable Anne Longley, Vancouver Police Department." HootSuite Social Media Management. Accessed August 2, 2013. http://blog. hootsuite.com/mydash-vpd/#utm_source=dlvr.it&utm_medium=twitter.

Lotan, Gilad, Erhardt Graeff, Mike Ananny, Devin Gaffney, Ian Pearce, and dana boyd. 2011. "The Revolutions Were Tweeted: Information Flows During the Tunisian and Egyptian Revolutions." *International Journal of Communication* 5:1375–405.

Lundby, Knut, ed. 2009. "Media Logic: Looking for Social Interaction." In *Mediatization: Concepts, Changes, Consequences*, edited by Knut Lundby, 101–22. New York: Peter Lang.

Macauley, Stuart. 1987. "Images of Law in Everyday Life: The Lessons of School, Entertainment and Spectator Sport." *Law and Society Review* 21:185–218.

Magid, Larry, and Anne Collier. 2007. *MySpace Unraveled: What It Is and How to Use It Safely*. Berkeley, CA: Peachpit Press.

Major Cities Chiefs Associates (MCCA), Major Counties Sheriffs Associates (MCSA), and Federal Bureau of Investigation National Executive Institute Associates (FBI-NEIA). 2013, July. *Social Media: A Valuable Tool with Risks*. MCCA, MCSA, and FBINEIA. http://www.neiassociates.org/storage/FBINEIA-2013SocMediaTool.pdf.

Major League Baseball. 2014. Replay Review Regulations. Accessed December 10. http://mlb.mlb.com/mlb/official_info/official_rules/replay_review.jsp.

Mandiberg, Michael. 2012. *The Social Media Reader*. New York: New York University Press.

Manning, Peter K. 1978. "The Police: Mandate, Strategies and Appearances." In *Policing: A View from the Street*, edited by P. K. Manning and J. van Maanen, 97–125. Santa Monica, CA: Goodyear.

———. 1992. *Organizational Communication*. Hawthorne, NY: Aldine Transaction.

———. 1997. *Police Work: The Social Organization of Policing*. 2nd ed. Prospect Heights, IL: Waveland Press.

———. 2003a. *Policing Contingencies*. Chicago: University of Chicago Press.

———. 2003b. *Symbolic Communication: Signifying Calls and the Police Response*. Cambridge, MA: MIT Press.

———. 2008. *The Technology of Policing: Crime Mapping, Information Technology, and the Rationality of Crime Control*. New York: New York University Press.

———. 2014. "Ethnographies of Policing." In *The Oxford Handbook of Police and Policing*, edited by Michael D. Reisig and Robert J. Kane, 518–47. New York: Oxford University Press.

———. 2015. "Researching Policing Using Qualitative Methods." In *The Routledge Handbook of Qualitative Criminology*, edited by Heith Copes and J. Mitchell Miller, 265–82. New York: Routledge.

Marfleet, Patti, Cindy John, Kim Russo, and Jane Devine. 1994. *Riots: A Background Paper: City of Vancouver Review of Major Events*. Vancouver: City of Vancouver. Accessed February 15, 2015. http://www.scribd.com/doc/58604466/Riots-A-Background-Paper-Part-1#scribd.

Marks, Monique. 2004. "Researching Police Transformation: The Ethnographic Imperative." *British Journal of Criminology* 44 (6): 866–88.

Marwick, Alice. 2008. "To Catch a Predator? The MySpace Moral Panic." *First Monday* 13 (6). http://firstmonday.org/article/view/2152/1966.

Marx, Gary. 1988. *Undercover: Police Surveillance in America*. Los Angeles: University of California Press.

Marx, Gary T., and James L. Wood. 1975. "Strands of Theory and Research in Collective Behavior." *Annual Review of Sociology* 1:363–428.

Mason, Gary. 2012. "One Year Later, Stanley Cup Riot a Distant Memory." *Globe and Mail*, June 15. http://www.theglobeandmail.com/news/british-columbia/one-year-laterstanley-cup-riot-a-distant-memory/article4277344/.

Mawby, Rob C. 1999. "Visibility, Transparency, and Police Media Relations." *Policing and Society* 9:263–86.

———. 2002a. "Continuity and Change, Convergence and Divergence: The Police and Practice of Police-Media Relations." *Criminal Justice* 2 (3): 303–24.

———. 2002b. *Policing Images: Policing, Communications and Legitimacy.* Cullompton, UK: Willan.

Mawby, Rob I. 2014. "Comparing Police Systems across the World." In *The Encyclopedia of Criminology and Criminal Justice*, edited by Gerben Bruinsma and David Weisburd, 478–88. New York: Springer.

McDiarmid, Jessica. 2013. "A City's Tolerance Ends: Outcry Grows as Sammy Yatim's Death Is Seen around the World." *Toronto Star*, July 31.

McGovern, Alyce. 2011. "Tweeting the News: Criminal Justice Agencies and Their Use of Social Networking Sites." Paper presented at the 2010 Australian and New Zealand Critical Criminology Conference, University of Sydney, Australia. http://ses.library.usyd.edu.au//bitstream/2123/7378/1/McGovern_ANZCCC2010.pdf.

McPhail, Clark. 1989. "Blumer's Theory of Collective Behavior: The Development of a Non-symbolic Interaction Explanation." *Sociological Quarterly* 30 (3): 401–23.

———. 1991. *The Myth of the Madding Crowd.* New York: Aldine de Gruyter.

———. 2006. "The Crowd and Collective Behavior: Bringing Symbolic Interaction Back In." *Symbolic Interaction* 4:433–64.

Mead, George H. 1934. *Mind, Self and Society.* Chicago: University of Chicago Press.

Mehta, Diana. 2015. "Toronto Police Star Year-Long Pilot Project to Test Body Cameras for Officers." *Globe and Mail*, May 15. http://www.theglobeandmail.com/news/toronto/toronto-police-start-year-long-pilot-project-to-test-body-cameras-for-officers/article24452340/.

Meier, Robert F. 1982. "Perspectives on the Concept of Social Control." *Annual Review of Sociology* 8:35–55.

Meijer, Albert, and Marcel Thaens. 2013. "Social Media Strategies: Understanding the Differences between North American Police Departments." *Government Information Quarterly* 30:343–50.

Mergel, Ines. 2010. "Government 2.0 Revisited: Social Media Strategies in the Public Sector." *Public Administration Times* 33 (3): 7–10.

Merringer, Ian. 2012. "What's the City Doing About Traffic Jams?; As It Turns Out, Not Much. Toronto Collects Volume Data, but When it Comes to Fixing the Congestion Problems, Its Hands Are Tied." *Globe and Mail*, March 10.

Metropolitan Police Interim Report: August Disorder. 2011. Metropolitan Police. November 30.

Meyrowitz, Joshua. 1985. *No Sense of Place.* New York: Oxford University Press.

Mezrich, Ben. 2010. *The Accidental Billionaires: The Founding of Facebook: A Tale of Sex, Money, Genius and Betrayal.* New York: Anchor.

Milbrandt, Tara. 2010. "On Appearing in Public in the 21st Century City: Ephemerality, Surveillance, and the Spector of the Visual Record." In *Cultural Production in Virtual and Imagined Worlds*, edited by Tracey Bowen and Mary Lou Nemanic, 115–35. Newcastle: Cambridge.

Milivojevic, Sanja, and Alyce McGovern. 2014. "The Death of Jill Meagher: Crime and Punishment on Social Media." *International Journal for Crime, Justice, and Social Democracy* 3 (3): 22–39.

Miller, Christa. 2012. "Now Tweeting: #copchat, the New Resource for Law Enforcement." *Cops 2.0*, June 20. http://cops2point0.com/2012/06/now-tweeting-copchat-new-resource-law-enforcement/.

Miller, Daniel. 2013. "Facebook's So Uncool, But It's Morphing into a Different Beast." *The Conversation*, December 19. http://theconversation.com/facebooks-so-uncool-but-its-morphing-into-a-different-beast-21548.

Miller, David. 2000. *Introduction to Collective Behavior and Collective Action.* Prospect Heights, IL: Waveland.

Miller, Vincent. 2011. *Understanding Digital Culture.* Thousand Oaks, CA: Sage.

Miller, W.R. 1975. "Police Authority in London and New York City 1830–1870." *Journal of Social History* 8 (2): 81–101.

Mills, C. Wright. 1940. "Situated Actions and Vocabularies of Motive." *American Sociological Review* 5 (6): 904–13.

Monahan, Torin, and Rodolfo D. Torres, eds. 2010. *Schools Under Surveillance: Cultures of Control in Public Education*. Piscataway, NJ: Rutgers University Press.

Muir, Jr., William K. 1977. *Police: Streetcorner Politicians*. Chicago: University of Chicago Press.

Murray, James. 2012. "Positive and Progressive Policing Includes Social Media and Public Engagement" *Netnewsledger*, January 1. http://www.netnewsledger.com/2012/01/01/positive-and-progressive-policing-includes-social-media-and-public-engagement/.

MySpace.com. 2006. *Law Enforcement Investigators Guide, 23 Jun 2006*. Published Wikileaks, December 6, 2009. http://wikileaks.org/wiki/MySpace.com_Law_Enforcement_Investigators_Guide,_23_Jun_2006.

Neal, David. M. 1993. "A Further Examination of Anonymity, Contagion, and Deindividuation in Crowds and Collective Behavior." *Sociological Focus* 26:93–107.

Newburn, Tim. 2015. "The 2011 England Riots in Recent Historical Perspective." *British Journal of Criminology* 55:39–64.

Newman, Maria. 2006. "MySpace.com Names Chief Security Officer." *New York Times*, April 11. http://www.nytimes.com/2006/04/11/technology/11cnd-myspace.html?pagewanted=all&_r=0.

Nolan, Kyle. 2014. "Rejecting the Frame: A Qualitative Analysis of Documented Police Violence in the Cases of Buddy Tavares and Robert Dziekanski." MA thesis, University of British Columbia–Okanagan, Kelowna, BC. https://circle.ubc.ca/bitstream/handle/2429/45987/ubc_2014_Fall_Nolan_Kyle.pdf?sequence=1.

Ontario Federation of Labour (OFL). 2013. "OFL Says the Toronto Police Services Boards' Failure to Follow Report Recommendations Led Directly to Death of Sammy Yatim." Press Release, July 31. http://ofl.ca/index.php/sammy-yatim/.

Opilo, Emily. 2010. "York City Police Reach out with Facebook Page." *York Daily Record* (Ontario), May 23.

O'Reilly, Tim. 2005. "What Is Web 2.0. O'Reilly Media, Inc." *O'Reilly Network*, September 30. http://www.evencone.net/wordpress2.8.1/wp-content/uploads/2010/01/OReillyNetwork_WhatIsWeb2.0.pdf.

Panagiotopoulos, Panagiotis, Alinaghi Ziaee Bigdeli, and Steven Sams. 2014. "Citizen-Government Collaboration on Social Media: The Case of Twitter in 2011 Riots in England." *Government Information Quarterly* 31:349-357.

Parascandola, Rocco. 2011. "NYPD Forms New Social Media Unit to Mine Facebook and Twitter for Mayhem." *New York Daily News*, August 10. http://www.nydailynews.com/new-york/nypd-forms-new-social-media-unit-facebook-twitter-mayhem-article-1.945242.

Park, Robert. 1927. "Human Nature and Collective Behavior." *American Journal of Sociology* 32 (5): 733–41.

Perinbanayagam, Robert. 1974. "The Definition of the Situation: An Analysis of the Ethnomethodological and Dramaturgical View." *Sociological Quarterly* 15 (4): 521–41.

Perverted Justice Foundation. 2015. "About Us." Perverted Justice Foundation Inc. Accessed November 17. http://www.pjfi.org/?pg=about.

Petrashek, Nathan. 2010. "The Fourth Amendment and the Brave New World of Online Social Networking." *Marquette Law Review* 93 (4): 1494–528.

Petrovic, Sasa, Miles Osborne, Richard McCreadie, Craig Macdonald, Iadh Ounis, and Luke Shrimpton. 2013. "Can Twitter Replace Newswire for Breaking News?" Proceedings of Seventh International Association for the Advancement of Artificial Intelligence (AAAI) Conference on Weblogs and Social Media. http://www.aaai.org/ocs/index.php/ICWSM/ICWSM13/paper/view/6066/6331.

Poisson, Jayme. 2012. "On the Social Media Beat: Const. Scott Mills is Breaking Rules, Pushing Boundaries—Even Saving Lives—Online." *Toronto Star*, March 10.

Police Services Act. R. S. O. 1990, c. P.15. http://www.ontario.ca/laws/statute/90p15.

Policing and Society. 2013. Special issue. "Policing Cybercrime: Networked and Social Media Technologies and Challenges for Policing" 23 (4).

Potts, Liza, and Angela Harrison. 2013. "Interfaces as Rhetorical Constructions: Reddit and 4chan during the Boston Marathon Bombings." Proceedings of the 31st ACM International Conference on Design of Communication. http://dl.acm.org/citation.cfm?id=2507079.

Prime Minister of Canada. 2013. Statement by the Prime Minister of Canada on National Victims of Crime Awareness week. Last modified April 21. *Government of Canada*. http://nouvelles.gc.ca/web/article-en.do?nid=734349.

Procter, Rob, Jeremy Crump, Susanne Karstedt, Alex Voss, and Marta Cantijoch. 2013a. "Reading the Riots: What *Were* Police Doing on Twitter?" *Policing and Society* 23 (4): 413–36.

———. 2013b. "Reading the Riots on Twitter: Methodological Innovation for the Analysis of Big Data." *International Journal of Social Research Methodology* 16 (3): 197–214.

Pue, Wes. 2000. *Pepper in Our Eyes: The APEC Affair*. Vancouver, BC: University of British Columbia Press.

Quickie, Simon. 2007. "The Next Generation." *MicroScope*, February 12.

Radford, Benjamin. 2006. "Predator Panic: A Closer Look." *Skeptical Inquirer* 30 (5). http://www.csicop.org/si/show/predator_panic_a_closer_look/.

Reiner, Robert. 2003. "Policing and the Media." In *Handbook of Policing*, edited by Tim Newburn, 259–81. Cullompton, UK: Willan.

———. 2010. *The Politics of the Police*. 4th ed. New York: Oxford University Press.

Reiss, Albert J. 1971. *The Police and the Public*. New Haven, CT: Yale University Press.

Reuters. 2013. "Former Police Chief Bratton Launches a Social Network for Cops." *CNBC*, June 29. http://www.cnbc.com/id/100854423.

Robertson, Ian. 2014. "Toronto Police Increasing Social Media Use." *Toronto Sun*, January 12. http://www.torontosun.com/2014/01/12/toronto-police-increasing-social-media-use.

Robinson, Laura. 2007. "The Cyberself: The Self-ing Project Goes Online, Symbolic Interaction in the Digital Age." *New Media & Society* 9 (1): 93–110.

Robinson, Matthew, Laura Kane, Evan Duggan, and Stephanie Law. 2011. "Vancouverites Fight Back Against Rioters Through Social Media." *Vancouver Sun*, June 17. http://www.vancouversun.com/news/Vancouverites+fight+back+against+rioters+through+social+media/4958109/story.html.

Romano, Andrew. 2006. "Walking a New Beat." *Newsweek*, April 26. http://www.newsweek.com/walking-new-beat-107825.

Rosenfeld, Michael J. 1997. "Celebration, Politics, Selective Looting and Riots: A Micro Level Study of the Bulls Riot of 1992 in Chicago." *Social Problems* 44 (4): 483–502.

Royal Canadian Mounted Police. (RCMP). 2014. "Important Notices." Accessed June 23, 2014. http://www.rcmp-grc.gc.ca/eng/important-notices.

Sacco, Vincent. 1995. "Media Constructions of Crime." *Annals of the American Academy of Political and Social Science* 539:141–54. http://www.umass.edu/legal/Benavides/Spring2005/397G/Readings%20397G%20Spring%202005/4Sacco.pdf.

Sammy's Fight Back for Justice Facebook page. 2013. Started July 28. https://www.facebook.com/SammysFightBackForJustice/info/?tab=page_info.

Scherer, Jay, and Lisa McDermott. 2012. "Hijacking Canadian Identity: Stephen Harper, Hockey, and The Terror of Neo-liberalism." In *Sport and Neo-liberalism: Politics, Consumption, and Culture*, edited by D. Andrews and M. Silk, 259–79. Philadelphia: Temple University Press.

Schneider, Christopher J. 2007. "Music and Media." *The Blackwell Encyclopedia of Sociology*. Vol. 6, edited by George Ritzer, 3129–34. Malden, MA: Oxford Blackwell Publishing.

———. 2009. "The Music Ringtone as an Identity Management Device." *Studies in Symbolic Interaction* 33:35–45.

———. 2011. "Constructing the Student Culprit." *Cultural Studies = Critical Methodologies* 11 (5): 434–45.

———. 2012. "American Crime Media in Canada: Law & Order and the Definition of the Situation." In *Present and Future of Symbolic Interactionism*, edited by A. Salvini, D. Altheide, and C. Nuti, 155–66. Pisa: Pisa University Press.

———. 2014. "Social Media and E-Public Sociology." In *The Public Sociology Debate: Ethics and Engagement*, edited by Ariane Hanemaayer and Christopher J. Schneider, 205–24. Vancouver, BC: University of British Columbia Press.

———. 2015a. "Meaning Making Online: Vancouver's 2011 Stanley Cup Riot." In *Kleine geheimnisse: Alltagssoziologische einsichten* (Little secrets: Everyday sociological insights), edited by Michael Dellwing, Scott Grills, and Heinz Bude, 81–102. Wiesbaden: Springer Germany.

———. 2015b. "Police Image Work in an Era of Social Media: YouTube and the 2007 Montebello Summit Protest." In *Social Media, Politics and the State: Protests, Revolutions, Riots, Crime and Policing in an Age of Facebook, Twitter and YouTube*, edited by Daniel Trottier and Christian Fuchs, 227–46. New York: Routledge.

———. 2015c. "Public Criminology and the 2011 Vancouver Riot: Public Perceptions of Crime and Justice in the 21st Century." *Radical Criminology* 5:21–46.

Schneider, Christopher J., and Daniel Trottier. 2012. "The 2011 Vancouver Riot and the Role of Facebook in Crowd-Sourced Policing." *BC Studies* 175 (Autumn): 93–109.

Schulz, Winfried. 2004. "Reconstructing Mediatization as an Analytical Concept." *European Journal of Communication* 19 (1): 87–101.

Schwartz, Daniel. 2013. "Why James Forcillo Was Charged with Murder in the Yatim Shooting." *CBC News*, August 20. http://www.cbc.ca/news/canada/why-james-forcillo-was-charged-with-murder-in-yatim-shooting-1.1386191.

Scott, Marvin B., and Stanford Lyman. 1968. "Accounts." *American Sociological Review* 33 (1): 46–62.

Scoville, Dean. 2009. "Watch What You Post: Social Networking Sites Are Great for Meeting New People and Having Some Fun, but Don't Let That Fun Kill Your Career." *Police Magazine*, December 10. http://www.policemag.com/channel/technology/articles/2009/12/watch-what-you-post.aspx.

Shaw, Gillian. 2012. "Fighting Crime, One Tweet at a Time; Police in B. C. Are Turning to Social Media to Engage with the Community, and As an Investigative Tool." *Vancouver Sun*, March 31.

Simmel, Georg. 1908. *Soziologie: Untersuchungen uber die Formen der Vergesellschaftung* (Sociology: Studies in the forms of sociation). Berlin: Duncker & Humblot.

———. 1950. *The Sociology of Georg Simmel*. Translated and edited by Kurt H. Wolff. New York: Free Press of Glencoe.

Skogan, Wesley G. 1990. *The Police and the Public in England and Wales: A British Crime Survey Report*. Home Office Research Study 117. London: HMSO.

———. 2005. "Citizen Satisfaction with Police Encounters." *Policing Quarterly* 8 (3): 298–321.

Skogan, Wesley G., and Kathleen Frydl. 2004. *Fairness and Effectiveness in Policing: The Evidence*. Washington, DC: National Academies Press.

Skogan, Wesley G., and Susan M. Hartnett. 1998. *Community Policing, Chicago Style*. New York: Oxford University Press.

Skolnick, Jerome, and James Fyfe. 1993. *Above the Law: Police and the Excessive Use of Force*. New York: Free Press.

Smelser, Neil J. 1962. *Theory of Collective Behavior*. Glencoe, IL: Free Press.

Smith, Craig. 2015. "By the Numbers: 17 MySpace Stats and Facts Then and Now." *DMR*, August 14. http://expandedramblings.com/index.php/myspace-stats-then-now/.

Snow, Robert P. 1983. *Creating Media Culture*. Newbury Park, CA: Sage.

Special Investigations Unit (SIU) (Ontario). 2015. "Frequently Asked Questions." Accessed December 9. http://www.siu.on.ca/en/faq.php.

Spiegelman, Sam. 2011. "Website Delivers Riot Justice in Vancouver; Posted Images Bust 'Morons.'" *USA Today*, July 8.

Sprague, Robert. 2007. "Fired for Blogging: Are There Legal Protections for Employees Who Blog?" *University of Pennsylvania Journal of Labor and Employment Law* 9 (2): 355–87.

Stark, Rodney. 1972. *Police Riots: Collective Violence and Law Enforcement*. Belmont, CA: Wadsworth.

Stebbins, Robert. 1967. "A Theory of the Definition of the Situation." *Canadian Review of Sociology and Anthropology* (Winter):148–64.

Stockwell, Jamie. 2005. "Missing Student's Online Musings Aid Search." *Washington Post*, October 3. http://www.washingtonpost.com/wp-dyn/content/article/2005/10/02/AR2005100201193.html.

Sumiala, Johanna, and Tikka Minttu. 2010. "'Web First' to Death: The Media Logic of the School Shootings in the Era of Uncertainty." *Nordicom Review: Nordic Research on Media & Communicaiton* 31:17–29. http://dialnet.unirioja.es/servlet/articulo?codigo=3509094.

Surette, Ray. 2010. *Media, Crime and Criminal Justice: Images, Realities and Polices*. 4th ed. Belmont, CA: Wadsworth.

S. W. 2013. "Why Do Germans Shun Twitter?" *Economist*, Babbage Science and Technology Blog, December 18. http://www.economist.com/blogs/babbage/2013/12/social-media.

Thomas, William Isaac, and Dorothy Thomas. 1928. *The Child in America: Behavior Problems and Programs*. New York: Knopf.

Thompson, John B. 1995. *The Media and Modernity: A Social Theory of the Media*. Cambridge: Polity.

———. 2000. *Political Scandal: Power and Visibility in the Media Age*. Cambridge: Polity.

———. 2005. "The New Visibility." *Theory, Culture & Society* 22 (6): 31–51.

Thorbes, Carol. 1994. "If You Know This Suspects, Touch Here." CBC Evening Television News, December 12. Accessed February 15, 2015. http://www.cbc.ca/archives/categories/society/crime-justice/the-long-lens-of-the-law/if-you-know-this-suspect-touch-here.html.

Tinati, Ramine, Susan Halford, Leslie Carr, and Catherine Pope. 2013. "The Promise of Bid Data: New Methods for Sociological Analysis." *The World Social Science Forum*. http://eprints.soton.ac.uk/358943/.

Tonkin, Emma, Heather D. Pfeiffer, and Greg Tourte. 2012. "Twitter, Information Sharing and the London Riots?" *Bulletin of the American Society for Information Science and Technology* 38:49–57.

Toronto Police Service (TPS). 2011a. "@torontopolice Preventing Crime." Accessed August 2, 2013. http://www.torontopolice.on.ca/modules.php?op=modload&name=News&file=article&sid=5808.

———. 2011b. "Service Launches Social Media Strategy." Retrieved August 2, 2013. http://www.torontopolice.on.ca/modules.php?op=modload&name=News&file=article&sid=5530.

———. 2011c. "Social Media Launch." YouTube. Uploaded July 29. www.youtube.com/watch?v=_BOpNtE0Gu4.

———. 2012. "Toronto Police Service Social Engagement Guidelines." http://www.torontopolice.on.ca/publications/files/social_media_guidelines.pdf.

———. 2013a. "Professional Standards." Accessed August 2. http://www.torontopolice.on.ca/professionalstandards/investigative.php.

———. 2013b. "Statement by Chief William Blair on Police Shooting." YouTube video, July 23. https://www.youtube.com/watch?v=9n9rUkbuiMg.

———. 2015. "Social Media." Accessed December 4. http://www.torontopolice.on.ca/socialmedia/.

Toronto Star. 2013a. "Man Shot in Streetcar Showdown with Police; Friends, Witnesses Shocked by Late-Night Confrontation That Left 18-Year-Old Dead." July 28.

———. 2013b. "Sammy Yatim: A Timeline." August 27. http://www.thestar.com/news/gta/2013/08/27/sammy_yatim_a_timeline.html.

Trottier, Daniel. 2012a. "Policing Social Media." *Canadian Review of Sociology* 49 (4): 411–25.

———. 2012b. *Social Media as Surveillance*. Farnham, UK: Ashgate.

———. 2014. *Identity Problems in the Facebook Era*. New York: Routledge.

———. 2015. "Vigilantism and Power Users: Police and User-Led Investigations on Social Media." In *Social Media, Politics and the State: Protests, Revolutions, Riots, Crime and Policing in an Age of Facebook, Twitter and YouTube*, edited by Daniel Trottier and Christian Fuchs, 209–26. New York: Routledge.

Tuchman, Gaye. 1978. *Making News: A Study in the Construction of Reality*. New York: Free Press.

Turner, Ralph H. 1994. "Race Riots Past and Present: A Cultural-Collective Behavior Approach." *Symbolic Interaction* 17 (3): 309–24.

Turner, Ralph H., and Lewis M. Killian. 1957. *Collective Behavior*. Englewood Cliffs, NJ: Prentice-Hall.

———. 1987. *Collective Behavior*. 3rd ed. Englewood Cliffs, NJ: Prentice-Hall.

Twitter. 2013. "#Twitter7." *The Twitter UK Blog*, March 21. https://blog.twitter.com/en-gb/2013/twitter7-0.

———. 2015. "Twitter Milestones: A Collection of Memorable Moments." Accessed December 4. https://about.twitter.com/company/press/milestones.

Tyler, Tom R. 2004. "Enhancing Police Legitimacy." *ANNALS of the American Academy of Political and Social Science* 593 (1): 84–99.

Tynan, Dan. 2006. "The 25 Worst Web Sites." *PC World*, September 15. http://www.pcworld.com/article/127116/article.html?page=7.

Vancouver Police Department (VPD). 2012. *2012–2016 Strategic Plan*. Vancouver, BC: Vancouver Police Department. http://vancouver.ca/police/assets/pdf/vpd-strategic-plan-2012-2016.pdf.

Vancouver Police Department, Planning, Research & Audit Section. 2012. *Administrative Report re: New Policy—Social Media 2.9.6 (iii) PR&A 2010-094*. Board Report #1224, February 28. http://vancouver.ca/police/policeboard/agenda/2012/0321/1224.pdf.

Vancouver Police Department (VPD), The Vancouver Riot Review Team. 2011. *Vancouver Police Department 2011 Stanley Cup Riot Review*. Vancouver, BC: Vancouver Police Department. http://vancouver.ca/police/assets/pdf/reports-policies/vpd-riot-review.pdf.

Van den Broeck, Tom. 2012. "Tides and Currents of Social Control: The Drift of Community Policing, a Belgian Case . . . " In *Tides and Currents of Policing Theories*, edited by Elke Devroe, Paul Ponsaers, Lodewijk Gunther Moor, Jack Greene, Layla Skinns, Lieselot Bisschop, Antoinette Verhage, and Matthew Bacon, 207–24. Antwerpen: Maklu-Publishers.

Van Dijck, Jose, and Thomas Poell. 2013. "Understanding Social Media Logic." *Media and Communication* 1 (1): 2–14.

Visser, Josh. 2013. "Sammy Yatim's Final Warning: New Audio Reveals Officer's Hostile Words before Teen Was Shot Dead by Police." *National Post*, July 29. http://news.nationalpost.com/2013/07/29/toronto-prepares-for-anti-police-demonstration-after-shooting-of-sammy-yatim-as-top-cop-promises-answers/.

Walker, Daniel. 1968. *Rights in Conflict. Convention Week in Chicago, August 25–29, 1968*. New York: E.P. Dutton.

Walking the Social Media Beat. 2015. "#copchat." Accessed November 24. http://walkingthesocialmediabeat.com/copchat/.

Waller, Willard. 1970. "The Definition of the Situation." In *Social Psychology Through Symbolic Interaction*, edited by G.P. Stone and H.A. Faberman, 162–74. Waltham, MA: Ginn-Blaidsdell.

Warmington, Joe. 2013. "Sammy Yatim's Facebook Page." *Toronto Sun*, August 2. http://www.torontosun.com/2013/08/02/quite-a-week-in-toronto.

Westley, William A. 1970. *Violence and the Police: A Sociological Study of Law, Custom, and Morality*. Cambridge, MA: MIT Press.

Williams, Alex. 2005. "Do You MySpace?" *New York Times*, August 28.

Williams, Kristian. 2007. *Our Enemies in Blue: Police and Power in America*. Cambridge, MA: South End Press.

Wintersgill, Carla. 2009. "Tweet! After the MMVA Party, a Celebrity Dust-Up." *Globe and Mail*. June 22. http://www.theglobeandmail.com/arts/tweet-after-the-mmva-party-a-celebrity-dust-up/article4277188/?from=1200833.

Wonneberger, A., K. Schoenbach, and L. van Meurs. 2013. "How Keeping up Diversifies: Watching Public Affairs TV in the Netherlands 1988–2010. " *European Journal of Communication* 28 (6): 646–62.

Wright, Sam. 1978. *Crowds and Riots: A Study in Social Organization*. Beverly Hills, CA: Sage.

Wünsch, Silke. 2012. "German Police Use Web 2.0 to Catch Criminals." *Deutsche Welle*, March 25. http://www.dw.de/german-police-use-web-20-to-catch-criminals/a-15836813.

York Regional Police (Ontario). 2014. "Media Relations Officers." Accessed April 30. http://www.yrp.ca/mediaofficers.aspx#sthash.o73P7GwA.dpuf.

Young, Kevin. 1986. "'The Killing Field': Themes in Mass Media Responses to the Heysel Stadium Riot." *International Review for the Sociology of Sport* 21:253–64.

Zhao, Shanyang, Sherri Grasmuck, and Jason Martin. 2008. "Identity Construction on Facebook: Digital Empowerment in Anchored Relationships." *Computers in Human Behavior* 24 (5): 1816–36.

Index

of 2011 riot, 56; investigation team findings on 2011 riot, 58; lack of social media use leading up to and during 2011 riot, 61–62, 63–64; and meaning making over riots, 7, 59; reaction to 2011 riot, xi, 58; reaction to Facebook's Vancouver Riot Pics page, 60; social media policy, 49, 82; use of Facebook, 30, 60, 63; use of social media in wake of 2011 riot, 62, 64–65, 74–76; use of Twitter, 61, 86

vigilante groups, 38, 60, 71, 73, 76
Virginia State Police, 51
Virji, Aly, 94
vocabularies of motive, 111

Web 2.0, 34
Wheet, Jody, 46
Wired Safety, 40

Yatim, Sammy: circumstances of his death, 99, 103; description of video of his death, 106–107; and discussion of the knife, 107–108, 108, 109–111, 111, 117, 119; Facebook profile of, 116–117; and Facebook's RIP Sammy Yatim page, 118, 119; and Facebook's Sammy's Fight Back for Justice page, 117–119; framing of video of, 107–109; news media's reporting of his death, 103, 104; and police use of force

question, 105, 111–112; posts on video of, 107, 110–111, 111–112; protests for, 105, 118, 119; social media's control of definition of his death, 102; and talk of mental illness, 111, 115, 117; TPS response to on social media, 112–114, 116, 121n8; and YouTube, xi, 9
Yatim, Sarah, 117
Yeo, Les, 58
York City Pennsylvania Police, 47
York Regional Police, 17, 18
Youth Criminal Justice Act (YCJA), 72–73
YouTube: and death of Sammy Yatim, 99, 103; and discussion of the knife in Sammy Yatim video, 107–108, 108, 109–111, 111; effect of video on view of Sammy Yatim's death, 9, 104, 111–112; framing of Sammy Yatim video, 107–109; popularity of video of Sammy Yatim's death, xi, 106–107; posts on Sammy Yatim video, 107, 110–111, 111–112; reaction to TPA, 115; and Sammy Yatim's Facebook profile on, 116; Toronto Police Association use of, 114, 115; TPS use of, 86, 116; and video clips on police brutality, 101; view of Special Investigations Unit on, 114

Zuckerberg, Mark, 44

About the Author

Christopher J. Schneider (PhD, Arizona State University) is associate professor of sociology at Brandon University in Manitoba, Canada. He taught at the University of British Columbia (UBC) for six years following the completion of his PhD in 2008 and for one year at Wilfrid Laurier University. His research investigates the role of media and information technology and related developments in social control. He has published three books and numerous academic journal articles and book chapters. His most recent co-edited book is *The Public Sociology Debate: Ethics and Engagement* (UBC Press, 2014). Schneider has received award recognition for his teaching, research, and service contributions. In 2013, he was the recipient of a Distinguished Academics Award, awarded by the Confederation of University Faculty Associations of British Columbia, which represents 4,600 university professors and other academic staff at the province's five doctoral universities. While at UBC, Schneider was the recipient of eight university-wide awards, including the Award for Teaching Excellence and Innovation, six consecutive Teaching Honour Roll Awards, and the Public Education Through Media Award "for actively and creatively sharing research expertise via the news media." Schneider is a frequent contributor to news media. His research and commentary have been featured in hundreds of news reports across North America, including the *New York Times*. Schneider has an English bulldog, likes baseball, and is an avid fan of the rock band Model Stranger.